LISREL® 8:

Structural Equation Modeling with the SIMPLIS™ Command Language

Karl G. Jöreskog and Dag Sörbom

Uppsala University

SSI SCIENTIFIC SOFTWARE INTERNATIONAL

Edited by Leo Stam and R. Darrell Bock

Cover by M. James Scott
Based on an architectural detail from Frank Lloyd Wright's Robie House.

ISBN: 0–89498–033–5

7 8 9 0   04 03 02 (fifth printing, with Foreword and computer exercises)

Published by:

Scientific Software International, Inc.
7383 North Lincoln Avenue, Suite 100
Lincolnwood, IL 60712–1704, USA
Tel: (847) 675–0720
Fax: (847) 675–2140
Web: *http://www.ssicentral.com*

Preface

This book introduces a new language for structural equation modeling. To use the new command language, all that is required is to name all observed and latent (if any) variables and to formulate the model to be estimated. The model can be specified either as paths or as relationships (equations) in the input file or as a path diagram at run time. It is not necessary to be familiar with the LISREL model or any of its submodels. No greek or matrix notations are required. There are no complicated options to learn. Anyone who can formulate the model as a path diagram can use the SIMPLIS command language immediately.

The book is written for students and researchers with limited mathematical and statistical training who need to use structural equation models to analyze their data and also for those who have tried but failed to learn the LISREL command language. It is not a textbook on factor analysis, structural equations or latent variable models although there are many examples of such in the book. Rather, it is assumed that the reader is already familiar with the basic ideas and principles of these types of analyses and techniques. A good introduction is Bollen's monograph *Structural equations with latent variables* (1989, Wiley). The objective of this book is to demonstrate that structural equation modeling can be done easily without all the technical jargon with which it has been associated. In a sense, the SIMPLIS language shifts the focus away from the technical question of *How to do it* so that researchers can concentrate on the more substantively interesting question *What does it all mean*? But, although, the SIMPLIS language makes it much easier to specify models and to carry out the analysis, the substantive specification and interpretation remain the same as with the LISREL command language.

This book is based on two principles:

☐ Learning by examples

☐ Learning by doing

Several examples of structural equation models are given in this book. The main objective of the examples is to illustrate the SIMPLIS command language. It is not to indicate the general applicability of LISREL nor to discuss the substantive interpretation of the output, although this is also done to some extent.

The SIMPLIS command language can be used with LISREL 8. LISREL 8 accepts two different command languages in the input file. To distinguish between these, we use the terms LISREL input and SIMPLIS input. A LISREL input is an input file written in the old LISREL command language as described in Jöreskog & Sörbom's *LISREL 7 - A guide to the program and applications* (1989, SPSS). A SIMPLIS input is an input file written in the new command language described in this book. One can use either the SIMPLIS language or the LISREL language in the input file but the two languages cannot be mixed in the same input file.

Beginning users of LISREL and users who often make mistakes when they specify the LISREL model will benefit greatly from using the SIMPLIS language, as this is much easier to learn and reduces the possibilities for mistakes to a minimum. Experienced users who seldom or never make mistakes in the model specification, may want to continue to use the LISREL command language.

This book describes the SIMPLIS command language and explains the SIMPLIS output obtained by default when LISREL 8 is used. It is also possible to obtain output in LISREL form even when the SIMPLIS language is used in the input file. For information on the LISREL command language and LISREL output, see the reference above or Jöreskog & Sörbom's *LISREL 7 User's Reference Guide* (1989, SSI).

The SIMPLIS command language can be learned from just a few examples. The general principles are easily understood intuitively from the examples. So we begin by giving some typical examples in Chapter 1. Examples of simultaneous analysis of data from several groups is covered in Chapter 2. Chapter 3 describes how one can obtain path diagrams for models on the screen and change them interactively at run time. The general issue of fitting and testing the model is discussed in Chapter 4.

LISREL output as distinguished from SIMPLIS output is described in Chapter 5. The general reference section for the SIMPLIS command language is given in Chapter 6.

Several persons have helped testing LISREL 8 and the SIMPLIS language. In particular, we like to thank Anne-Marie Aish and Yang Fan, who did a more thorough testing than most others.

Karl G. Jöreskog
Dag Sörbom
Uppsala, Sweden, January 1993

Contents

LIST OF EXAMPLES   xv

EDITORS' FOREWORD   xxi

1   SIMPLE EXAMPLES   1
   1.1   Regression Models   1
      1.1.1   A Single Regression Equation   1
          Example 1: Regression of GNP   2
      1.1.2   Bivariate Regression   6
          Example 2: Prediction of Grade Averages   6
   1.2   Path Analysis   11
          Example 3: Union Sentiment of Textile Workers   12
   1.3   Measurement Models   15
          Example 4: Ability and Aspiration   16
   1.4   Confirmatory Factor Analysis   22
          Example 5: Nine Psychological Variables - A Confirmatory Factor Analysis   23
   1.5   Path Analysis with Latent variables   28
      1.5.1   Recursive System   29
          Example 6: Stability of Alienation   29
          Example 7: Performance and Satisfaction   34
      1.5.2   Non-Recursive System   38
          Example 8: Peer Influences on Ambition   38
   1.6   Analysis of Ordinal variables   44
          Example 9: Panel Model for Political Efficacy   45

2   MULTI-SAMPLE EXAMPLES   51
   2.1   Equal Factor Structures   52

Example 10: Testing Equality of Factor Structures   52

Example 11: Parental Socioeconomic Characteristics   56

2.2   Equal Regressions   61

Example 12: Testing Equality of Regressions . .   61

2.3   Estimation of Means of Latent Variables   66

Example 13: Mean Difference in Verbal Ability .   66

Example 14: Nine Psychological Variables with Factor Means   71

2.4   Regression Models with Latent Variables   77

Example 15: Regression of Verbal7 on Verbal5 . .   77

Example 16: Head Start Summer Program . . .   79

3   PATH DIAGRAMS   **85**

3.1   Parameter Estimates and t-Values   86

3.2   B-, X-, Y-, S-, and R-Diagrams   87

3.3   Fit Statistics   92

3.4   Modification Indices   93

3.5   Adding or Deleting a Path   94

3.6   Adding an Error Covariance   96

3.7   Relaxing an Equality Constraint   97

3.8   Starting with an Empty Path Diagram   100

3.9   Zooming   104

3.10 Printing a Path Diagram   105

3.11 Saving the Path Diagram   106

3.12 Significance Levels for t-Values and Modification Indices .   107

3.12.1 t-Values   107

3.12.2 Modification Indices   108

3.13 Summary of Keys   108

4   FITTING AND TESTING   **111**

4.1   From Theory to Statistical Model   111

4.2   Nature of Inference   114

4.3   Fitting and Testing a Covariance Structure   115

4.4   Selection of One of Several a priori Specified Models . .   118

4.5   Model Assessment and Modification   120

4.5.1   Chi-square   121

4.5.2   Other Goodness-of-Fit Measures   122

Goodness-of-Fit Indices   122

Fit Measures Based on Population Error of Approximation 123

Information Measures of Fit 124

Other Fit Indices 125

4.5.3 Detailed Assessment of Fit 126

Fitted and Standardized Residuals 126

Modification Index 127

4.5.4 Strategy of Analysis 128

4.6 Illustration 129

5 LISREL OUTPUT **133**

5.1 Hypothetical Model 133

5.2 Classification of Variables 136

5.3 Parameter Matrices 137

5.4 Parameter Specifications 140

5.5 LISREL Estimates 141

5.6 Goodness-of-Fit Statistics 144

5.7 Fitted and Standardized Residuals 146

5.8 Modification Indices 147

5.9 Standardized Solutions 151

5.10 Direct, Indirect, and Total Effects (EF) 154

5.11 Estimating the Standardized Solution Directly 158

6 SIMPLIS REFERENCE **161**

6.1 Input File 161

6.2 Title 162

6.3 Observed Variables or Labels 163

6.4 Data 165

6.4.1 Raw Data 165

6.4.2 Covariance Matrix or Correlation Matrix . . . 166

6.4.3 Means and Standard Deviations 169

6.4.4 Asymptotic Covariance Matrix 169

6.4.5 Asymptotic Variances 170

6.4.6 Selection of Variables 170

6.5 Sample Size 170

6.6 Latent Variables or Unobserved Variables 171

6.7 Relationships 171

6.8 Paths 172

6.9 Scaling the Latent Variables 173
6.10 Starting Values 174
6.11 Error Variances and Covariances 175
    6.11.1 Fixed Error Variances 175
    6.11.2 Error Covariances 175
6.12 Uncorrelated Factors 176
6.13 Equality Constraints 177
    6.13.1 Equal Paths 177
    6.13.2 Equal Error Variances 177
    6.13.3 Freeing a Fixed Parameter or Relaxing an Equality
    Constraint 178
6.14 Options 178
    6.14.1 Wide Print 179
    6.14.2 Print Residuals 179
    6.14.3 Number of Decimals 179
    6.14.4 Method of Estimation 180
    6.14.5 Admissibility Check 181
    6.14.6 Maximum Number of Iterations 182
    6.14.7 Save Sigma 183
6.15 Cross-Validation 183
6.16 LISREL Output 184
6.17 End of Problem 185

7 COMPUTER EXERCISES   187
Exercise 1 187
    Problems 188
Exercise 2 188
    Problem A 189
    Problem B 190
    Problem C 190
Exercise 3 192
    Problems 193
Exercise 4 193
    Problems 196
Exercise 5 196
    Problem A 196
    Problem B 199

Exercise 6 199
    Problem 200
Exercise 7 201
    Problem A 203
    Problem B 203
    Problem C 203
Exercise 8 203
    Problem A 204
    Problem B 205
    Problem C 205
    Problem D 206
    Problem E 206
Exercise 9 209
    Problems 209

REFERENCES   **211**

AUTHOR INDEX   **219**

SUBJECT INDEX   **221**

x

List of Tables

1.1 Covariance Matrix for GNP Data   3

1.2 Scores for Fifteen College Freshmen on Five Educational Measures   7

1.3 Covariance Matrix for Union Sentiment Variables . . .   14

1.4 Correlations Among Ability and Aspiration Measures . .   16

1.5 Correlation Matrix for Nine Psychological Variables . .   24

1.6 Covariance Matrix for Stability of Alienation   30

1.7 Means, Standard Deviations and Correlations for the Observed Variables in Bagozzi's Model   36

1.8 Parameter Estimates with and without Measurement Error in VERBINTM   38

1.9 Correlations and Standard Deviations for Background and Aspiration Measures for 329 Respondents and Their Best Friends   41

1.10 Loadings and their Standard Errors   49

1.11 Error Variances, Reliabilities and Autocovariances . .   50

1.12 Correlations between Efficacy and Responsiveness . .   50

1.13 Stability Coefficients and Residual Covariance Matrix .   50

2.1 Covariance Matrices for SAT Verbal and Math Sections .   52

2.2 Covariance Matrices for Parental Socioeconomic Characteristics   57

2.3 Estimated Reliabilities of Son's and Parents' Reports of Parental Socioeconomic Characteristics   61

2.4 Means and Covariance Matrices for STEP Reading and Writing   62

2.5 Estimated Means and Covariance Matrices of Verbal5 and Verbal7   70

2.6 Nine Psychological Variables: Correlations 72
2.7 Nine Psychological Variables: Means and Standard Deviations 73
2.8 Maximum Likelihood Estimates for Nine Psychological Variables with Factor Means 76
2.9 Correlations, Standard Deviations and Means for the Head Start Data 80

7.1 Correlations for Background, Aspiration, and Attainment Variables 187
7.2 Covariance Matrix for Role Behavior Variables 190
7.3 Covariance Matrix for Split-halves Variables 191
7.4 Covariance Matrix for Two Versions of Television Possession and Watching at Three Occasions 193
7.5 Covariance Matrix of Control of Pace and Powerlessness Measured by Three Methods at Two Points in Time . . 195
7.6 Correlation Matrix for Nine Psychological Tests . . . 199
7.7 Correlations for Variables in MIMIC Model 200
7.8 Covariance Matrices for Five Variables on Perceived and Subjective Socioeconomic Status for Blacks and Whites . 201
7.9 Correlations, Means and Standard Deviations for Indicators of Objective Class and Subjective Class. 204

List of Figures

1.1 Path Diagram for Regression of GNP 2
1.2 Path Diagram for Prediction of Grade Averages . . . 8
1.3 Path Diagram for Union Sentiment Model 12
1.4 Path Diagram for Ability and Aspiration 17
1.5 Confirmatory Factor Analysis Model for Nine Psychological Variables 25
1.6 Model for Stability of Alienation 31
1.7 Modified Model for Performance and Satisfaction . . . 35
1.8 Path Diagram for Peer Influences on Ambition 40
1.9 Panel Model for Political Efficacy 47

2.1 Path Diagram for SAT Verbal and Math 53
2.2 Path Diagram for Parental Socioeconomic Characteristics 58
2.3 Path Diagram for Regression of READING7 63
2.4 Path Diagram for Estimating Mean of Verbal5 67
2.5 Path Diagram for Estimating the Mean of Verbal5 and Verbal7 69
2.6 Confidence Ellipses for Verbal5 and Verbal7 71
2.7 Factor Mean Profiles 77
2.8 Head Start: Model for Problem C 83

3.1 Diagram with Parameter Estimates for EX1A.SPL . . 86
3.2 Diagram with t-values for EX1A.SPL 87
3.3 B-diagram with Estimates for EX6B.SPL 88
3.4 S-diagram with Estimates for EX6B.SPL 89
3.5 X-diagram with Estimates for EX6B.SPL 90
3.6 Y-diagram with Estimates for EX6B.SPL 91
3.7 R-diagram with Estimates for EX6B.SPL 92

3.8 B-diagram with Modification Indices for EX5A.SPL . . 93

3.9 R-diagram with Modification Indices for EX9A.SPL . . 97

3.10 B-diagram with Estimates of Group 1 for EX10C.SPL . 98

3.11 B-diagram with Estimates of Group 2 for EX10C.SPL . 99

3.12 Empty B-diagram for EX3B.SPL 101

3.13 B-diagram for EX3B.SPL with Paths Drawn 102

3.14 Empty B-diagram for EX4B.SPL 103

3.15 B-diagram for EX6C.SPL with Paths Drawn 105

5.1 Path Diagram for Hypothetical Model 134

5.2 Path Diagram in Greek Notation 139

5.3 Reciprocal Causation Between Eta1 and Eta2 154

7.1 Model for Educational and Occupational Aspiration . . 188

7.2 Longitudinal Model for Two Versions of Television Possession and Watching at Three Occasions 194

7.3 Confirmatory Factor Analysis Model for Nine Psychological Tests 197

7.4 Second-Order Factor Analysis Model for Nine Psychological Tests 198

7.5 Path Diagram for MIMIC Model 200

7.6 Model for Objective and Subjective Socioeconomic Status 202

7.7 Model A: Measurement Model for Objective Class . . . 207

7.8 Model B: Measurement Model for Subjective Class . . 207

7.9 Model C: Structural Equation Model for Objective and Subjective Class 208

7.10 Model D: Mimic Model for Objective and Subjective Class 208

List of Examples
(with Input and Data Files)

This book has many examples illustrating most of the common types of models and methods used with LISREL. For beginners of LISREL it is instructive to go over these examples to learn how to set up the input file for particular models and problems. We also suggest using these examples as exercises in the following ways:

- Estimate the same model with a different method of estimation
- Estimate the same model from correlations instead of covariances or vice versa
- Request other options for the output
- Formulate and test hypotheses about the parameters of the model
- Estimate a different model for the same data
- Make deliberate mistakes in the input file and see what happens

Input and data files for these examples are available on diskette. For these files we use the following naming conventions.

The first letters in the file refer to the example in the book. Thus, EX13B means Example 13B. Input files have the suffix .SPL for SIMPLIS. The suffix after the period in the name of the data file refers to the type of data it contains:

- LAB for labels
- COV for covariance matrix
- COR for correlation matrix
- RAW for raw data
- DAT for a file containing several types of data

- □ PML for matrix of polychoric (and polyserial) correlations produced by PRELIS under listwise deletion
- □ KML for matrix of product-moment correlations (based on raw scores or normal scores) produced by PRELIS under listwise deletion
- □ ACP for asymptotic covariance matrix of the elements of a PML matrix produced by PRELIS
- □ ACK for asymptotic covariance matrix of the elements of a KML matrix produced by PRELIS

Example 1: Regression of GNP   2–6, 86–86
            Data File: None
            Input Files: EX1A.SPL, EX1B.SPL

Example 2: Prediction of Grade Averages   6–10
            Data File: None
            Input Files: EX2A.SPL, EX2B.SPL

Example 3: Union Sentiment of Textile Workers   12–15, 100–101
            Data File: None
            Input Files: EX3A.SPL, EX3B.SPL

Example 4: Ability and Aspiration   16–21, 101–102
            Data File: EX4.COR
            Input Files: EX4A.SPL, EX4B.SPL

Example 5: Nine Psychological Variables   23–28, 92–95, 129–131
            Data File: EX5.COR
            Input Files: EX5A.SPL, EX5B.SPL

Example 6: Stability of Alienation   29–34, 87–91, 103–104
            Data File: None
            Input Files: EX6A.SPL, EX6B.SPL, EX6C.SPL

Example 7: Performance and Satisfaction   34–38
            Data File: EX7.DAT
            Input Files: EX7A.SPL, EX7B.SPL

Example 8: Peer Influences on Ambition   38–44
            Data Files: EX8.LAB, EX8.COR, EX8.STD
            Input Files: EX8A.LS7, EX8B.LS7, EX8C.LS7,
                         EX8D.LS7

Example 9: Panel Model for Political Efficacy   45–49, 96–97
            Data Files: PANEL.LAB, PANELUSA.PME,
                         PANELUSA.ACP
            Input Files: EX9A.SPL, EX9B.SPL

Example 10: Testing Equality of Factor Structures   52–56, 97–99
             Data File: EX10.COV
             Input Files: EX10A.SPL, EX10B.SPL, EX10C.SPL,
                          EX10D.SPL

Example 11: Parental Socioeconomic Characteristics   56–60
             Data File: None
             Input Files: EX11A.SPL, EX11B.SPL

Example 12: Testing Equality of Regressions   61–66
             Data File: EX12.DAT
             Input Files: EX12A.SPL, EX12B.SPL, EX12C.SPL

Example 13: Mean Difference in Verbal Ability   66–70
             Data File: EX12.DAT
                          (Note: Same Data File as for Example 12)
             Input Files: EX13A.SPL, EX13B.SPL

Example 14: Nine Psychological Variables with Factor Means   71–77
             Data Files: EX14.LAB, EX14.DAT
             Input File: EX14.SPL

Example 15: Regression of Verbal7 on Verbal5   77–79
             Data File: EX12.DAT
                          (Note: Same Data File as for Example 12)
             Input Files: EX15A.SPL, EX15B.SPL

Example 16: Head Start Summer Program   79–84
             Data File: EX16.DAT
             Input Files: EX16A.SPL, EX16B.SPL, EX16C.SPL,
                          EX16D.SPL

Example 17: Hypothetical Model   133–159
             Data Files: EX17.COV, EX17.COR
             Input Files: EX17A.SPL, EX17B.SPL

Editors' Foreword

Most first, and even second, courses in applied statistics seldom go much further than ordinary least squares analysis of data from controlled experiments, group comparisons, or simple prediction studies. Collectively, these procedures make up regression analysis, and the linear mathematical functions on which they depend are referred to as regression models. This basic method of data analysis is quite suitable for curve-fitting problems in physical science, where an empirical relationship between an observed dependent variable and a manipulated independent variable must be estimated. It also serves well the purposes of biological investigation in which organisms are assigned randomly to treatment conditions and differences in the average responses among the treatment groups are estimated.

An essential feature of these applications is that only the dependent variable or the observed response is assumed to be subject to measurement error or other uncontrolled variation. That is, there is only one random variable in the picture. The independent variable or treatment level is assumed to be fixed by the experimenter at known predetermined values. The only exception to this formulation is the empirical prediction problem. For that purpose, the investigator observes certain values of one or more predictor variables and wishes to estimate the mean and variance of the distribution of a criterion variable among respondents with given values of the predictors. Because the prediction is conditional on these known values, they may be considered fixed quantities in the regression model. An example is predicting the height that a child will attain at maturity from his or her current height and the known heights of the parents. Even though all of the heights are measured subject to error, on the childs height at maturity is considered a random variable.

Where ordinary regression methods no longer suffice, and indeed give misleading results, is in purely observational studies in which all variables are subject to measurement error or uncontrolled variation and the purpose of the inquiry is to estimate relationships that account for variation among the variables in question. This is the essential problem of data analysis in those fields where experimentation is impossible or impractical and mere empirical prediction is not the objective of the study. It is typical of almost all research in fields such as sociology, economics, ecology, and even areas of physical science such as geology and meteorology. In these fields, the essential problem of data analysis is the estimation of structural relationships between quantitative observed variables. When the mathematical model that represents these relationships is linear we speak of a *linear* structural relationship. The various aspects of formulating, fitting, and testing such relationships we refer to as *structural equation modeling*.

Although structural equation modeling has become a prominent form of data analysis only in the last twenty years (thanks in part to the availability of the LISREL program), the concept was first introduced nearly eighty years ago by the population biologist, Sewell Wright, at the University of Chicago. He showed that linear relationships among observed variables could be represented in the form of so-called *path diagrams* and associated *path coefficients*. By tracing causal and associational paths on the diagram according to simple rules, he was able to write down immediately the linear structural relationship between the variables. Wright applied this technique initially to calculate the correlation expected between observed characteristics of related persons on the supposition of Mendelian inheritance. Later, he applied it to more general types of relationships among persons.

The modern form of linear structural analysis includes an algebraic formulation of the model in addition to the path diagram representation. The two forms are equivalent and the implementation of the analysis in the LISREL 8 program permits the user to submit the model to the computer in either representation. The path analytic approach is excellent when the number of variables involved in the relationship is moderate, but the diagram becomes cumbersome when the number of variables is large. In that case, writing the relationships symbolically is more convenient. This text presents examples of both representations and makes clear the correspondence between the paths and the structural equations.

Notice that in the above mentioned fields in which experimentation is hardly ever possible, psychology and education do not appear. Controlled experiments with both animal and human subjects have been a mainstay of psychological research for more than a century, and in the 1920s experimental evaluations of instructional methods began to appear in education. As empirical research developed in these fields, however, a new type of data analytic problem became apparent that was not encountered in other fields.

In psychology, the difficulty was, and still is, that for the most part there are no well-defined dependent variables. The variables of interest differ widely from one area of psychological research to another and often go in and out of favor within areas over relatively short periods of time. Psychology has been variously described as the science of behavior or the science of human information processing. But the varieties of behavior and information handling are so multifarious that no progress in research can be made until investigators identify the variables to be studied and the method of observing them. Where headway has been made in defining a coherent domain of observation, it has been through the mediation of a construct—some putative latent variable that is modified by stimuli from various sources and in turn controls or influences various observable aspects of behavior. The archetypal example of such a latent variable is the construct of general intelligence introduced by Charles Spearman to account for the observed positive correlations between successful performance on many types of problem-solving tasks.

Investigation of mathematical and statistical methods required in validating constructs and measuring their influence led to the development of the data analytic procedure called *factor analysis*. Its modern form is due largely to the work of Truman Kelly and L.L. Thurstone, who transformed Spearman's one-factor analysis into a fully general multiple-factor analysis. More recently, Karl Jöreskog added confirmatory factor analysis to the earlier exploratory form of analysis. The two forms serve different purposes. Exploratory factor analysis is an authentic discovery procedure: it enables one to see relationships among variables that are not at all obvious in the original data or even in the correlations among variables. Confirmatory factor analysis enables one to test whether relationships expected on theoretical grounds actually appear in the data. Derrick Lawley and Karl Jöreskog provided a statistical procedure, based on maximum likelihood estimation, for fitting factor models to data and testing the number of factors that can be detected and reliably estimated in the data.

Similar problems of defining variables appear in educational research, even in experimental studies of alternative methods of instruction. The goals of education are broad and the outcomes of instruction are correspondingly many: an innovation in instructional practice may lead to a gain in some measured outcomes and a loss in others. The investigator can measure a great many such outcomes, but unless all are favorable or all unfavorable the results become too complex to discuss or provide any guide to educational policy. Again, factor analysis is a great assistance in identifying the main dimensions of variation among outcomes and suggesting parsimonious constructs for their discussion.

In the LISREL model, the linear structural relationship and the factor structure are combined into one comprehensive model applicable to observational studies in many fields. The model allows 1) multiple latent constructs indicated by observable explanatory (or *exogenous*) variables, 2) recursive and nonrecursive relationships between constructs, and 3) multiple latent constructs indicated by observable responses (or *endogenous*) variables. The connections between the latent constructs compose the structural equation model; the relationships between the latent constructs and their observable indicators or outcomes compose the factor models. All parts of the comprehensive model may be represented in the path diagram and all factor loadings and structural relationships appear as coefficients of the path.

Nested within the general model are simpler models that the user of the LISREL program may choose as special cases. If some of the variables involved in the structural relationships are observed directly, rather than indicated, part or all of the factor model may be excluded. Conversely, if there are no structural relationships, the model may reduce to a confirmatory factor analysis applicable to the data in question. Finally, if the data arise from a simple prediction problem or controlled experiment in which the independent variable or treatment level is measured without error, the user may specialize to a simple regression model and obtain the standard results of ordinary least-squares analysis.

These specializations may be communicated to the LISREL 8 computer program in three different ways. At the most intuitive, visual level, the user may construct the path diagram interactively on the screen, and specify paths to be included or excluded. The corresponding verbal level is the SIMPLIS command language described in this text. It requires only that the user name the variables and declare the relationships among them. The third and most detailed level is the LISREL command language. It is phrased in terms of matrices that appear in the matrix-algebraic representation of the model. Various parameters of the matrices may be fixed or set equal to other parameters, and linear and non-linear constraints may be imposed among them. The terms and syntax of the LISREL command language are explained and illustrated in the LISREL 8 program manuals. Most but not all of these functions are included in the SIMPLIS language; certain advanced functions are only possible in native LISREL commands.

The essential statistical assumption of LISREL analysis is that random quantities within the model are distributed in a form belonging to the family of elliptical distributions, the most prominent member of which is the multivariate normal distribution. In applications where it is reasonable to assume multivariate normality, the maximum likelihood method of estimating unknowns in the model is justified and usually preferred. Where the requirements of maximum likelihood estimation are not met, as when the data are ordinal rather than measured, the various least squares estimation methods are available. It is important to understand, however, except in those cases where ordinary least squares analysis applies or the weight matrices of other least squares methods are known, that these are large-sample estimation procedures. This is not a serious limitation in observation studies, where samples are typically large. Small-sample theory applies properly only to controlled experiments and only when the model contains a single, univariate or multivariate normal error component.

The great merit of restricting the analytical methods to elliptically distributed variation is the fact that the sample mean and covariance matrix (or correlation matrix and standard deviations) are sufficient statistics of the analysis. This allows all the information in the data that bear on the choice and fitting of the model to be compressed into the relatively small number of summary statistics. The resulting data compression is a tremendous advantage in large-scale sample-survey studies, where the number of observations may run to the tens of thousands, whereas the number of sufficient statistics are of an order of magnitude determined by the number of variables.

The operation of reducing the raw data to their sufficient statistics (while cleaning and verifying the validity for the data) is performed by the PRELIS program which accompanies LISREL 8. PRELIS also computes summary statistics for qualitative data in the form of tetrachoric or polychoric correlation matrices. When there are several sample groups, and the LISREL model is defined and compared across the groups, PRELIS prepares the sufficient statistics for each sample in turn.

In many social and psychological or educational research studies where a single sample is involved, the variables are ususally measured on a scale with an arbitrary origin. In that case, the overall means of the variables in the sample can be excluded from the analysis, and the fitting of the LISREL model can be regarded simply as an analysis of the covariance structure, in which case the expected covariance matrix implied by the model is fitted to the observed covariance matrix directly. Since the sample covariance matrix is a sufficient statistic under the distribution assumption, the result is equivalent to fitting the data. Again, the analysis is made more manageable because one can examine the residuals from the observed covariances, which are moderate in number, as opposed to analyzing residuals of the original observations in a large sample.

Many of these aspects of the LISREL analysis are brought out in the examples in this text and in the LISREL 8 program manual. In addition, the present text contains exercises to help the student strengthen and expand his or her understanding of this powerful method of data analysis. Files containing the data of these examples are included with the program and can be analyzed in numerous different ways to explore and test alternative models.

READINGS

Bollen, K.A. (1989) *Structural equations with latent variables.* New York: Wiley.

Jöreskog, K.G., & Sörbom, D. (1989) *LISREL 7 User's Reference Guide.* Chicago: Scientific Software International.

Jöreskog, K.G., & Sörbom, D. (1993) *New features in LISREL 8.* Chicago: Scientific Software International.

1 SIMPLE EXAMPLES

This chapter introduces the SIMPLIS command language by means of simple examples. Models for directly observed variables without measurement errors are introduced first. These include regression models and recursive path models discussed in Sections 1 and 2. Measurement models and confirmatory factor analysis models involving observed indicators of latent variables are considered in Sections 3 and 4. Various path models for latent variables are illustrated in the remaining sections. An example with ordinal variables is used in Section 6.

1.1 Regression Models

1.1.1 A Single Regression Equation

A regression equation

$$y = \alpha + \gamma_1 x_1 + \gamma_2 x_2 + \cdots + \gamma_q x_q + z \qquad (1.1)$$

is used to describe the linear relationship between a dependent variable y and a set of independent (explanatory, causal, or predictor) variables $x_1, x_2, \ldots, x_q$. The random error term z is assumed to be uncorrelated with the independent variables. The x-variables may be fixed or random variables. If they are fixed, it is assumed that the distribution of z does not depend on the values of $x_1, x_2, \ldots, x_q$.

The error term z is an aggregate of all variables influencing y, but not included in the relationship. Using another terminology, z is called a *stochastic disturbance* term in the equation.

The uncorrelatedness between z and the x-variables is a crucial assumption. Studies should be planned and designed so that this assumption is met. Failure to do so may lead to considerable bias in estimated γ-coefficients. This is sometimes called *omitted variables bias*.

Regression analysis is one of the most widely used techniques in the social as well as other sciences; see, e.g., Johnston (1972), Goldberger (1964), and Draper & Smith (1967). It is usually employed to estimate the regression equation and the relative explanatory power of each x-variable or to determine the best predictors of the dependent variable y.

As an example of how to use LISREL to estimate a single regression equation, consider the following example.

Example 1: Regression of GNP

Goldberger (1964, p. 187) presented raw data on gross national product in billions of dollars (y = GNP), labor inputs in millions of man-years (x_1 = LABOR), real capital in billions of dollars (x_2 = CAPITAL), and the time in years measured from 1928 (x_3 = TIME). A path diagram for the regression of y on $x_1, x_2,$ and x_3 is shown in Figure 1.1. The data consists of 23 annual observations for the United States during 1929–1941 and 1946–1955. The covariance matrix of the variables is given in Table 1.1.

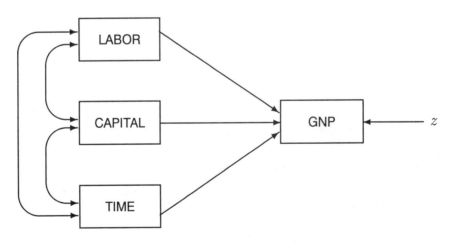

Figure 1.1   Path Diagram for Regression of GNP

As the path diagram indicates, LABOR, CAPITAL, and TIME are supposed to influence GNP. This is indicated by the three one-way (unidirected) arrows pointing towards GNP. The three two-way arrows on the left indicate that the three independent variables may be correlated. The one-way arrow on the right represents the effect of the error term z.

The means and the covariance matrix may be computed using the PRELIS program, which can take missing values and other data problems into account.

The SIMPLIS input file (**EX1A.SPL**):

```
Regression of GNP
Observed variables: GNP LABOR CAPITAL TIME
Covariance Matrix
 4256.530
  449.016     52.984
 1535.097    139.449   1114.447
  537.482     53.291    170.024     73.747
Sample Size: 23
Equation: GNP = LABOR CAPITAL TIME
End of Problem
```

The first line is a title line. Any number of title lines may be used and one may write anything on the title lines as long as they do not begin with the words *Observed Variables* or *Labels*. For further rules, see Section 6.2 of Chapter 6.

The second line defines the names of the variables. The order of the names of the variables must correspond to the order of the variables in the covariance matrix.

The following five lines define the covariance matrix. Only the lower half of the covariance matrix is given. The elements of the covariance matrix are given in free format, i.e., with blanks between them.

Table 1.1   Covariance Matrix for GNP Data

	y	x_1	x_2	x_3
GNP	4256.530			
LABOR	449.016	52.984		
CAPITAL	1535.097	139.449	1114.447	
TIME	537.482	53.291	170.024	73.747
Means	180.435	45.565	50.087	13.739

The next line specifies the sample size, i.e., the number of cases on which the covariance matrix is based.

The line

```
Equation: GNP = LABOR CAPITAL TIME
```

specifies the regression equation to be estimated and is interpreted as "GNP depends on LABOR, CAPITAL, and TIME". This may also be specified as

```
Paths: LABOR - TIME -> GNP
```

which is interpreted "There is a path from each of the variables LABOR, CAPITAL, and TIME to GNP".

The line End of Problem, which is optional, may be used to specify the end of the problem. In this case, this is also the end of the input file.

The estimated regression equation shows up in the output file as:

```
GNP = 3.82*LABOR + 0.32*CAPITAL + 3.79*TIME, Errorvar.= 12.47, R² = 1.00
      (0.22)        (0.031)          (0.19)                 (4.05)
       17.70         10.54            20.35                  3.08
```

The estimated regression coefficients appear in front of the * before each variable. The estimated partial regression coefficient of LABOR is 3.82. This is interpreted as follows. If LABOR increases one unit while CAPITAL and TIME are held fixed, the expected increase of GNP is 3.82 units on the average.

The error variance is 12.47, which is small compared to the total variance 4256.53 of GNP. This indicates that the three independent variables LABOR, CAPITAL, and TIME account for almost all of the variance of GNP.

The numbers below the regression coefficients are the standard errors of the estimates. Each standard error is a measure of the precision of the parameter estimate. Below the standard errors are the t-values. The t-value is the ratio between the estimate and its standard error. If a t-value exceeds a certain level, we say that the corresponding parameter is *significant* which means that one can be fairly confident that the corresponding variable really influences GNP. In small samples, if the residuals are normally distributed, one can use a formal F test to test whether a specified regression coefficient is zero in the population. In our example, all the regression coefficients are highly significant.

The squared multiple correlation, R^2 is also given for each relationship. This is a measure of the strength of the linear relationship. A formal test of the significance of the whole regression equation, i.e., a test of the hypothesis that all γ's are zero, can be obtained by computing

$$F = \frac{R^2/q}{(1 - R^2)/(N - q - 1)} \, , \tag{1.2}$$

where R^2 is the squared multiple correlation listed in the output file, N is the total sample size and q is the number of genuine x-variables. F is used as an F-statistic with q and $N - q - 1$ degrees of freedom. In our example, $R^2 = 0.997$ and $F = 2104.8$ with 3 and 19 degrees of freedom. Because the program uses two decimals by default, 0.997 is rounded up to 1.00. To get three decimals in the output, include the line

```
Number of Decimals = 3
```

in the input file (see Section 6.14.3 of Chapter 6).

Observe that the regression model (1.1) is scale-invariant in the following sense. If y is replaced by $y^* = c_0 y$ and x_i replaced by $x_i^* = c_i x_i, i = 1, 2, \ldots, q$, where the c's are arbitrary non-zero constants, the analysis of these scaled variables will yield regression coefficients

$$\hat{\gamma}_i^* = (c_0/c_i)\hat{\gamma}_i.$$

The standard error will change similarly, but the t-values are invariant under such scalings of the variables.

The constant or intercept term α in (1.1) is the mean of y when all x-variables are zero. When only the covariance matrix is given in the input file, LISREL 8 assumes that all variables are measured in deviations from their means and that α is zero. To estimate α, all one needs to do is to include the means of the variables in the input file.

The input file for estimating α is (**EX1B.SPL**):

```
Regression of GNP
Observed variables: GNP LABOR CAPITAL TIME
Means: 180.435 45.565 50.087 13.739
Covariance Matrix:
 4256.530
  449.016      52.984
 1535.097     139.449    1114.447
```

```
   537.482     53.291    170.024      73.747
Sample Size: 23
Equation: GNP = LABOR CAPITAL TIME
End of Problem                                                        ,
```

The estimated regression equation will then be

```
GNP =  - 61.73 + 3.82*LABOR + 0.32*CAPITAL + 3.79*TIME, Errorvar.= 12.47,
         (7.77)  (0.22)       (0.031)        (0.19)                 (4.05)
         -7.94   17.70         10.54          20.35                  3.08
```

$$R^2 = 1.00$$

Note that the estimated regression parameters and their standard errors are the same as in Example 1A. The intercept term is estimated at -61.73 with a standard error of 7.77. This indicates that GNP would be highly negative if LABOR, CAPITAL, and TIME were all zero. This is an extrapolation which assumes that the relationship is linear over the entire range of values of the x-variables. More interesting examples of intercept terms in equations are given in Chapter 2.

Analysis of variance (ANOVA) and analysis of covariance (ANCOVA) can also be done with regression analysis by including dummy variables in the regression equation (see Huitema, 1980, or Jöreskog & Sörbom, 1989, pp. 112–116).

1.1.2 Bivariate Regression

The following example illustrates the case of two dependent variables y_1 and y_2, and three explanatory variables x_1, x_2, and x_3.

Example 2: Prediction of Grade Averages

Finn (1974) presents the data given in Table 1.2. These data represent the scores of fifteen freshmen at a large midwestern university on five educational measures. The five measures are:

$y_1 = $ *grade average for required courses taken (GRAVEREQ)*

$y_2 = $ *grade average for elective courses taken (GRAVELEC)*

$x_1 = $ *high-school general knowledge test, taken previous year (KNOW-LEDG)*

Table 1.2
Scores for Fifteen College Freshmen on Five Educational Measures

Case	y_1	y_2	x_1	x_2	x_3
1	.8	2.0	72	114	17.3
2	2.2	2.2	78	117	17.6
3	1.6	2.0	84	117	15.0
4	2.6	3.7	95	120	18.0
5	2.7	3.2	88	117	18.7
6	2.1	3.2	83	123	17.9
7	3.1	3.7	92	118	17.3
8	3.0	3.1	86	114	18.1
9	3.2	2.6	88	114	16.0
10	2.6	3.2	80	115	16.4
11	2.7	2.8	87	114	17.6
12	3.0	2.4	94	112	19.5
13	1.6	1.4	73	115	12.7
14	.9	1.0	80	111	17.0
15	1.9	1.2	83	112	16.1

$x_2 = IQ$ *score from previous year (IQPREVYR)*

$x_3 = educational$ *motivation score from previous year (ED MOTIV)*

We examine the predictive value of x_1, x_2 and x_3 in predicting the grade averages y_1 and y_2. The path diagram of the model is shown in Figure 1.2.

The SIMPLIS input for such a model is extremely simple (**EX2A.SPL**):

```
Prediction of Grade Averages
Observed Variables: GRAVEREQ GRAVELEC KNOWLEDG IQPREVYR 'ED MOTIV'
Raw Data
0.8 2.0 72 114 17.3
2.2 2.2 78 117 17.6
1.6 2.0 84 117 15.0
2.6 3.7 95 120 18.0
2.7 3.2 88 117 18.7
2.1 3.2 83 123 17.9
```

```
3.1 3.7 92 118 17.3
3.0 3.1 86 114 18.1
3.2 2.6 88 114 16.0
2.6 3.2 80 115 16.4
2.7 2.8 87 114 17.6
3.0 2.4 94 112 19.5
1.6 1.4 73 115 12.7
0.9 1.0 80 111 17.0
1.9 1.2 83 112 16.1
Relationships
    GRAVEREQ = KNOWLEDG  IQPREVYR  'ED MOTIV'
    GRAVELEC = KNOWLEDG  IQPREVYR  'ED MOTIV'
End of Problem
```

The variables are listed by labels. Note that since the label for the last variable includes a blank space, it must be enclosed within single quotes.

The data for this example is raw data which is read in free format with one line per case. The sample size need not be specified. The program determines the sample size by counting the number of lines of data. In applications where the raw data contain many cases, it is recommended that the raw data be stored in a data file rather than in the input file.

The relationships specify that there are two equations to be estimated, one for each dependent variable. Each equation is specified by listing the dependent variable first and then the three independent variables.

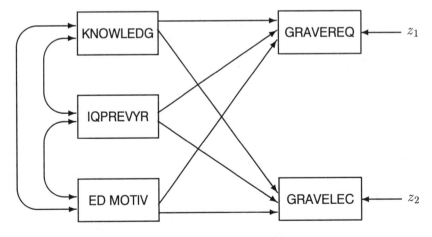

Figure 1.2   Path Diagram for Prediction of Grade Averages

The three right-hand variables in each equation are consecutive variables, so one can replace the second right-hand variable by a hyphen and write the two relationships:

```
GRAVEREQ = KNOWLEDG - 'ED MOTIV'
GRAVELEC = KNOWLEDG - 'ED MOTIV'
```

The two relationships can be simplified even further by recognizing that the two left-hand variables are also consecutive variables and the same right-hand variables are involved in each relationship. The two relationships can therefore be specified on a single line:

```
GRAVEREQ - GRAVELEC = KNOWLEDG - 'ED MOTIV'
```

This is interpreted by the program as "Each of the variables from GRAVEREQ to GRAVELEC depends on all of the variables from KNOWLEDG to ED MOTIV."

Alternatively, one can specify the model by paths:

```
KNOWLEDG - 'ED MOTIV' -> GRAVEREQ - GRAVELEC
```

which is short for

```
KNOWLEDG - 'ED MOTIV' -> GRAVEREQ
KNOWLEDG - 'ED MOTIV' -> GRAVELEC
```

The two relationships are estimated as

```
GRAVEREQ =  - 5.62  + 0.085*KNOWLEDG + 0.0082*IQPREVYR - 0.015*ED MOTIV,
            (5.61)  (0.027)          (0.049)          (0.11)
            -1.00    3.17             0.17             -0.13

            Errorvar.= 0.26 , R² = 0.57
                      (0.11)
                       2.35

GRAVELEC =  - 20.40 + 0.047*KNOWLEDG + 0.15*IQPREVYR + 0.13*ED MOTIV,
            (5.40)  (0.026)          (0.047)         (0.11)
            -3.78    1.82             3.12            1.17

            Errorvar.= 0.24 , R² = 0.69
                      (0.10)
                       2.35
```

The t-values show that only KNOWLEDG is a significant predictor of GRAVEREQ, and only IQPREVYR is a significant predictor of GRAV-ELEC. ED MOTIV is not significant for either purpose. It should be emphasized, however, that the sample size is too small to draw any safe conclusions.

The output also contains a section called **GOODNESS OF FIT STATISTICS**, the first line of which is:

```
CHI-SQUARE WITH 1 DEGREE OF FREEDOM = 8.89 (P = 0.0029)
```

LISREL 8 assumes by default that the two error terms z_1 and z_2 are uncorrelated. This assumption means that y_1 and y_2 would be uncorrelated after the effects of x_1, x_2, and x_3 are removed. The chi-square value 8.89 given in the output file is a formal test of this assumption. Since it is significant, one may be interested in estimating the error covariance, i.e., the covariance between z_1 and z_2. This can be done by including the following line in the input file (see file **EX2B.SPL**):

```
Let the Error Terms of GRAVEREQ and GRAVELEC be Correlated
```

Alternatively, this may be specified as:

```
Set the Error Covariance between GRAVEREQ and GRAVELEC Free
```

The output file for the model with correlated error terms shows the same estimated relationships and then gives the estimated covariance matrix of the two error terms z_1 and z_2:

```
Error Covariance for GRAVELEC and GRAVEREQ = 0.17
                                           (0.090)
                                             1.88
```

The error covariance is 0.17. The partial correlation between y_1 and y_2 for given x_1, x_2, x_3 is $0.17/\sqrt{(0.26 \times 0.24)} = 0.68$.

Suppose we want to test the hypothesis that ED MOTIV has no effect on neither GRAVEREQ nor GRAVELEC. One might think that this is just a matter of deleting ED MOTIV from the two relationships in the input file **EX2B.SPL**. However, this will not work because the resulting model is still saturated and fits the data perfectly. So no test of the hypothesis is provided by the program.

To test the hypothesis, one should instead specify the relationships as

```
GRAVEREQ = KNOWLEDG   IQPREVYR   0*'ED MOTIV'
GRAVELEC = KNOWLEDG   IQPREVYR   0*'ED MOTIV'
```

The resulting output gives a chi-square of 3.40 with two degrees of freedom. So the hypothesis is not rejected.

With the SIMPLIS command language, selection of variables is automatic in the sense that only the variables involved in the model are used even if there are more variables in the data (see Section 6.4.6). Thus, when ED MOTIV is deleted from the relationships, the program uses only the first four variables and estimates the two regressions of GRAVEREQ and GRAVELEC on KNOWLEDG and IQPREVYR, and with correlated error terms, this is a saturated model. When the relationships are specified as above, ED MOTIV is still in the model but its effect on GRAVEREQ and GRAVELEC is forced to be zero.

1.2 Path Analysis

Path analysis, due to Wright (1934), is a technique to assess the direct causal contribution of one variable to another in a nonexperimental situation. The problem, in general, is that of estimating the coefficients of a set of linear *structural equations*, representing the cause and effect relationships hypothesized by the investigator. The system involves variables of two kinds: independent or cause variables $x_1, x_2, \ldots, x_q$ and dependent or effect variables $y_1, y_2, \ldots, y_p$. The classical technique consists of first solving the structural equations for the dependent variables in terms of the independent variables and the random disturbance terms $z_1, z_2, \ldots, z_p$ to obtain the *reduced form equations*, estimating the regression of the dependent variables on the independent variables and then solving for the structural parameters in terms of the regression coefficients. The last step is not always possible. Models of this kind and a variety of estimation techniques have been extensively studied by econometricians (see Theil, 1971); by biometricians (see Turner & Stevens, 1959, and references therein); and by sociologists (see Blalock, 1985, and Duncan, 1975). Some of these models involve latent variables; see Duncan (1966), Werts & Linn (1970), and Hauser & Goldberger (1971).

To begin, we shall consider path analysis models for directly observed variables. Estimating a path analysis model for directly observed variables with LISREL 8 is straightforward. Rather than estimating each equation separately, LISREL 8 considers the model as a system of equations and estimates all the structural coefficients directly. The reduced form is obtained as a by-product.

The fundamental difference between this type of model and a regression model is that dependent variables appear also on the right side of the relationships. The following example is a *recursive system* in the sense that the dependent variables can be ordered in a sequence such that each dependent variable depends only on x-variables and previous dependent variables.

Example 3: Union Sentiment of Textile Workers

McDonald & Clelland (1984) analyzed data on union sentiment of southern nonunion textile workers. After transformation of one variable and treatment of outliers, Bollen (1989a) reanalyzed a subset of the variables according to the model shown in Figure 1.3. The variables are:

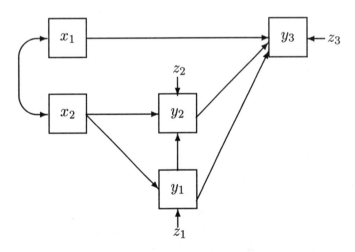

Figure 1.3   Path Diagram for Union Sentiment Model

$y_1 =$ *deference (submissiveness) to managers*
$y_2 =$ *support for labor activism*
$y_3 =$ *sentiment towards unions*
$x_1 =$ *logarithm of years in textile mill*
$x_2 =$ *age*

The covariance matrix is given in Table 1.3.

In the two previous examples, we have used names of variables like LA-BOR, CAPITAL, and TIME in Example 1, and GRAVEREQ and IQPRE-VYR in Example 2. While this is a recommended practice, some users may prefer to use generic names such as Y1, Y2, Y3, X1, and X2 as in Figure 1.3.

The model in Figure 1.3 can be easily specified as relationships (equations):

```
Relationships
Y1 = X2
Y2 = X2 Y1
Y3 = X1 Y1 Y2
```

These three lines express the following three statements, respectively:

- Y1 depends on X2
- Y2 depends on X2 and Y1
- Y3 depends on X1, Y1, and Y2

It is also easy to formulate the model by specifying its paths:

```
Paths
X1 -> Y3
X2 -> Y1 Y2
Y1 -> Y2 Y3
Y2 -> Y3
```

These lines express the following statements:

- There is one path from X1 to Y3
- There are two paths from X2, one to Y1 and one to Y2
- There are two paths from Y1, one to Y2 and one to Y3
- There is one path from Y2 to Y3

Altogether there are six paths in the model corresponding to the six one-way (unidirected) arrows in Figure 1.3.

Table 1.3   Covariance Matrix for Union Sentiment Variables

y_1	y_2	y_3	x_1	x_2
14.610				
−5.250	11.017			
−8.057	11.087	31.971		
−0.482	0.677	1.559	1.021	
−18.857	17.861	28.250	7.139	215.662

The input file for this problem is straightforward (file **EX3A.SPL**):

```
Title: Union Sentiment of Textile Workers

        Variables: Y1 = deference (submissiveness) to managers
                   Y2 = support for labor activism
                   Y3 = sentiment towards unions
                   X1 = Years in textile mill
                   X2 = age

Observed Variables: Y1 - Y3 X1 X2
Covariance Matrix:
   14.610
   -5.250  11.017
   -8.057  11.087  31.971
   -0.482   0.677   1.559   1.021
  -18.857  17.861  28.250   7.139 215.662
Sample Size 173
Relationships
   Y1 = X2
   Y2 = X2 Y1
   Y3 = X1 Y1 Y2
End of Problem
```

The first nine lines are title lines that define the problem and the variables. To indicate the beginning of the title lines, one may use the word Title, although this is not necessary. The first real command line begins with the words Observed Variables. Note that Y1, Y2, and Y3 may be labeled collectively as Y1-Y3 (see Section 6.3 of Chapter 6). The observed variables, the covariance matrix, and the sample size are defined as in Example 1. The model is specified as relationships.

The model makes a fundamental distinction between two kinds of variables: dependent and independent. The *dependent variables* are supposed to be explained by the model, i.e., variation and covariation in the dependent variables are supposed to be accounted for by the *independent variables*. The dependent variables are on the left side of the equal sign. They correspond to those variables in the path diagram which have one-way arrows pointing towards them. Variables which are not dependent are called independent. The distinction between dependent and independent variables is already inherent in the labels: x-variables are independent and y-variables are dependent. Note that y-variables can appear on the right side of the equations.

The output gives the following estimated relationships:

```
Y1 =  - 0.087*X2, Errorvar.= 12.96, R² = 0.11
        (0.019)               (1.41)
        -4.65                 9.22


Y2 =  - 0.28*Y1 + 0.058*X2, Errorvar.= 8.49 , R² = 0.23
        (0.062)   (0.016)              (0.92)
        -4.58     3.59                 9.22


Y3 =  - 0.22*Y1 + 0.85*Y2 + 0.86*X1, Errorvar.= 19.45, R² = 0.39
        (0.098)   (0.11)    (0.34)              (2.11)
        -2.23     7.53      2.52                9.22
```

1.3 Measurement Models

Broadly speaking, there are two basic problems that are important in the social and behavioral sciences. The first problem is concerned with the measurement properties—validities and reliabilities—of the measurement instruments. The second problem concerns the causal relationships among the variables and their relative explanatory power.

Most theories and models in the social and behavioral sciences are formulated in terms of theoretical or hypothetical concepts, or constructs, or latent variables, which are not directly measurable or observable. However, often a number of indicators or symptoms of these variables can be used to represent the latent variables more or less well. The purpose of a measurement model is to describe how well the observed indicators serve as a measurement instrument for the latent variables. The key concepts here are measurement, reliability, and validity. Measurement models often suggest ways in which the observed measurements can be improved.

Measurement models are important in the social and behavioral sciences when one tries to measure such abstractions as people's behavior, attitudes, feelings and motivations. Most measures employed for such purposes contain sizable measurement errors and the measurement models allow us to take these errors into account.

Example 4: Ability and Aspiration

Calsyn & Kenny (1977) presented the correlation matrix in Table 1.4 based on 556 white eighth-grade students. The measures are:

$x_1 = $ *self-concept of ability (S-C ABIL)*

$x_2 = $ *perceived parental evaluation (PPAREVAL)*

$x_3 = $ *perceived teacher evaluation (PTEAEVAL)*

$x_4 = $ *perceived friend's evaluation (PFRIEVAL)*

$x_5 = $ *educational aspiration (EDUC ASP)*

$x_6 = $ *college plans (COL PLAN)*

Table 1.4   Correlations Among Ability and Aspiration Measures

	x_1	x_2	x_3	x_4	x_5	x_6
S-C ABIL	1.00					
PPAREVAL	0.73	1.00				
PTEAEVAL	0.70	0.68	1.00			
PFRIEVAL	0.58	0.61	0.57	1.00		
EDUC ASP	0.46	0.43	0.40	0.37	1.00	
COL PLAN	0.56	0.52	0.48	0.41	0.72	1.00

We analyze a model in which x_1, x_2, x_3, and x_4 are assumed to be indicators of "ability" and x_5 and x_6 are assumed to be indicators of "aspiration." We are primarily interested in estimating the correlation between true ability and true aspiration. The path diagram for this example is given in Figure 1.4. To distinguish latent variables from observed variables in the path diagram, the former are enclosed in ovals and the latter in rectangles.

In this example we illustrate the use of an external data file. The data file contains the correlation matrix of the variables. The name of the file is **EX4.COR**. The file is as follows.

```
1.00
0.73 1.00
0.70 0.68 1.00
0.58 0.61 0.57 1.00
0.46 0.43 0.40 0.37 1.00
0.56 0.52 0.48 0.41 0.72 1.00
```

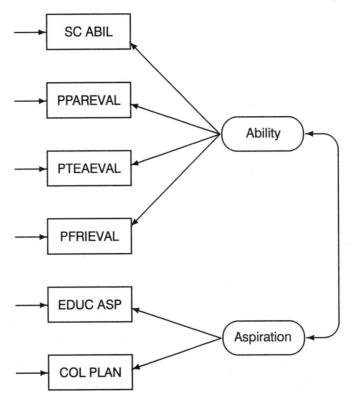

Figure 1.4   Path Diagram for Ability and Aspiration

The input file illustrates, in particular, how to specify the model by paths instead of by relationships (file **EX4A.SPL**).

```
Ability and Aspiration
Observed Variables
'S-C ABIL' PPAREVAL PTEAEVAL PFRIEVAL 'EDUC ASP' 'COL PLAN'
Correlation Matrix From File: EX4.COR
Sample Size: 556
Latent Variables: Ability Aspiratn
Paths
Ability  -> 'S-C ABIL' PPAREVAL PTEAEVAL PFRIEVAL
Aspiratn -> 'EDUC ASP' 'COL PLAN'
Print Residuals
End of Problem
```

Lines 2 and 3 define the observed variables as before. Line 4 specifies that the correlation matrix is to be read from the file **EX4.COR** rather than from the input file. Line 5 gives the sample size. This model involves latent (unobservable) variables as well as observed variables, so a distinction must be made between these two kinds of variables in the input file. The latent variables are defined in line 6. In this book we name latent variables using capitalized words to distinguish them from observed variables whose names will be entirely in capitals. The model is specified in terms of paths. The first line of the paths section (line 8) specifies that there is a path from Ability to each of the variables from S-C ABIL to PFRIEVAL. This implies 4 paths. The next line (line 9) specifies that there is a path from Aspiration to both EDUC ASP and COL PLAN. This means 2 more paths. The line `Print Residuals` is a request to print certain matrices of residuals. This will be explained later.

The output file reveals the following results.

$$\text{S-C ABIL} = 0.86 * \text{Ability}, \quad \text{Errorvar.} = 0.25 \quad , \ R^2 = 0.75$$
$$\quad\quad\quad (0.035) \quad\quad\quad\quad\quad\quad (0.023)$$
$$\quad\quad\quad 24.56 \quad\quad\quad\quad\quad\quad\quad 10.91$$

$$\text{PPAREVAL} = 0.85 * \text{Ability}, \quad \text{Errorvar.} = 0.28 \quad , \ R^2 = 0.72$$
$$\quad\quad\quad (0.035) \quad\quad\quad\quad\quad\quad (0.024)$$
$$\quad\quad\quad 23.96 \quad\quad\quad\quad\quad\quad\quad 11.55$$

$$\text{PTEAEVAL} = 0.81 * \text{Ability}, \quad \text{Errorvar.} = 0.35 \quad , \ R^2 = 0.65$$
$$\quad\quad\quad (0.036) \quad\quad\quad\quad\quad\quad (0.027)$$
$$\quad\quad\quad 22.11 \quad\quad\quad\quad\quad\quad\quad 13.07$$

```
PFRIEVAL = 0.70*Ability, Errorvar.= 0.52  , R² = 0.48
           (0.039)                (0.035)
           18.00                  14.88


EDUC ASP = 0.78*Aspiratn, Errorvar.= 0.40  , R² = 0.60
           (0.040)                (0.038)
           19.21                  10.45


COL PLAN = 0.93*Aspiratn, Errorvar.= 0.14  , R² = 0.86
           (0.039)                (0.044)
           23.57                  3.15
```

CORRELATION MATRIX OF INDEPENDENT VARIABLES

```
               Ability   Aspiratn
               --------  --------
Ability          1.00

Aspiratn          .67      1.00
                 (.03)
                 21.53
```

The latent variables are standardized by default unless some other unit of measurement is specified by the user (see Section 6.9 of Chapter 6). Since the observed variables are also standardized in this example, this means that the factor loadings which appear in front of each latent variable above are *standardized* factor loadings or standardized validity coefficients (see Bollen, 1989a).

The correlation between Ability and Aspiration is estimated as .67 with a standard error of .03 and t-value of 21.53. The high t-value indicates that the correlation is non-zero. In another context, it may be more interesting to test whether the correlation is 1. This can be done by forming an approximate confidence interval for the true correlation using the standard error. In this case, the confidence interval will be (.604, .728). Since this interval does not include the value 1, we conclude that the correlation is less than 1. This is not a rigorous test, but it gives a crude method of testing whether a uni-dimensional model rather than a two-dimensional model would be sufficient to account for the intercorrelations among the observed variables.

The correlation between Ability and Aspiration is a disattenuated correlation in the sense that the effects of measurement errors in the observed variables have been eliminated. The disattenuated correlation of

.67 between Ability and Aspiration may be compared with the attenuated correlations between the observed ability and aspiration measures, which range between .37 and .56. Thus the disattenuated correlation is higher.

For measurement relations, there is an error term on each observed variable as indicated by the one-way arrows on the left side of the path diagram. These error terms represent *errors in variables* rather than *errors in equations* as the error terms in the previous examples. The error terms are usually interpreted as measurement errors (or observational errors) in the observed variables, although they may also contain specific systematic components. The estimated error variance is given along with each measurement relationship. The squared multiple correlation R^2 is also given for each equation. This is a measure of the strength of the linear relationship. In this context, R^2 is usually interpreted as the reliability of the observed measure on the left. It is seen that S C ABIL is the most reliable of the indicators of Ability and COL PLAN is the most reliable indicator of Aspiration. Since the model fits the data well, we may also interpret the loadings in front of the latent variables as validity coefficients and interpret S C ABIL as the most valid indicator of Ability and COL PLAN as the most valid indicator of Aspiration.

The fit of the model is quite good as evidenced by the chi-square of 9.26 with 8 degrees of freedom. As explained in Section 4.5 of Chapter 4, the fit may be examined further by inspecting the sections in the output file labeled **FITTED COVARIANCE MATRIX**, **FITTED RESIDUALS**, and **STANDARDIZED RESIDUALS**. These look as follows.

```
        FITTED COVARIANCE MATRIX

            S-C ABIL    PPAREVAL    PTEAEVAL    PFRIEVAL    EDUC ASP    COL PLAN
  S-C ABIL     1.00
  PPAREVAL      .73        1.00
  PTEAEVAL      .69         .68        1.00
  PFRIEVAL      .60         .59         .56        1.00
  EDUC ASP      .45         .44         .42         .36        1.00
  COL PLAN      .53         .53         .50         .43         .72        1.00


        FITTED RESIDUALS

            S-C ABIL    PPAREVAL    PTEAEVAL    PFRIEVAL    EDUC ASP    COL PLAN
  S-C ABIL      .00
  PPAREVAL      .00         .00
  PTEAEVAL      .01         .00         .00
```

```
PFRIEVAL       -.02         .02         .01         .00
EDUC ASP        .01        -.01        -.02         .01         .00
COL PLAN        .03        -.01        -.02        -.02         .00         .00
```

```
         STANDARDIZED RESIDUALS

            S-C ABIL   PPAREVAL   PTEAEVAL   PFRIEVAL   EDUC ASP   COL PLAN
S-C ABIL       .00
PPAREVAL      -.61         .00
PTEAEVAL       .73        -.50         .00
PFRIEVAL     -1.91        1.69         .72         .00
EDUC ASP      1.01        -.57        -.87         .46         .00
COL PLAN      2.09        -.43       -1.13        -.95         .00         .00
```

If the line

```
Print Residuals
```

is not included in the input file, the information about the residuals is only given in summarized form as follows:

```
SUMMARY STATISTICS FOR FITTED RESIDUALS
SMALLEST FITTED RESIDUAL =      -.02
  MEDIAN FITTED RESIDUAL =       .00
 LARGEST FITTED RESIDUAL =       .03


STEMLEAF PLOT
 - 2|00
 - 1|86
 - 0|96430000000
   0|5
   1|0149
   2|6


SUMMARY STATISTICS FOR STANDARDIZED RESIDUALS
SMALLEST STANDARDIZED RESIDUAL =    -1.91
  MEDIAN STANDARDIZED RESIDUAL =      .00
 LARGEST STANDARDIZED RESIDUAL =     2.09


STEMLEAF PLOT
 - 1|91
 - 0|9966540000000
   0|577
   1|07
   2|1
```

1.4   Confirmatory Factor Analysis

It is important to distinguish between exploratory and confirmatory analysis. In an exploratory analysis, one wants to explore the empirical data to discover and detect characteristic features and interesting relationships without imposing any definite model on the data. An exploratory analysis may be structure generating, model generating, or hypothesis generating. In confirmatory analysis, on the other hand, one builds a model assumed to describe, explain, or account for the empirical data in terms of relatively few parameters. The model is based on *a priori* information about the data structure in the form of a specified theory or hypothesis, a given classificatory design for items or subtests according to objective features of content and format, known experimental conditions, or knowledge from previous studies based on extensive data.

Exploratory factor analysis is a technique often used to detect and assess latent sources of variation and covariation in observed measurements. It is widely recognized that exploratory factor analysis can be quite useful in the early stages of experimentation or test development. Thurstone's (1938) primary mental abilities, French's (1951) factors in aptitude and achievement tests and Guilford's (1956) structure of intelligence are good examples of this. The results of an exploratory analysis may have heuristic and suggestive value and may generate hypotheses which are capable of more objective testing by other multivariate methods. As more knowledge is gained about the nature of social and psychological measurements, however, exploratory factor analysis may not be a useful tool and may even become a hindrance.

Most studies are to some extent both exploratory and confirmatory since they involve some variables of known and other variables of unknown composition. The former should be chosen with great care in order that as much information as possible about the latter may be extracted. It is highly desirable that a hypothesis which has been suggested by mainly exploratory procedures should subsequently be confirmed, or disproved, by obtaining new data and subjecting these to more rigorous statistical techniques. Although LISREL is most useful in confirmatory studies, it can also be used to do exploratory analysis by means of a sequence of confirmatory analyses. It must be emphasized, however, that one must have at least a tentative theory or hypothesis to start out with.

The basic idea of factor analysis is the following. For a given set of response variables $x_1, \ldots, x_q$ one wants to find a set of underlying latent factors $\xi_1, \ldots, \xi_n$, fewer in number than the observed variables. These latent factors are supposed to account for the intercorrelations of the response variables in the sense that when the factors are partialed out from the observed variables, there should no longer remain any correlations between these. This leads to the model (see Jöreskog, 1979a)

$$x_i = \lambda_{i1}\xi_1 + \lambda_{i2}\xi_2 + \cdots + \lambda_{in}\xi_n + \delta_i \,, \qquad (1.3)$$

where δ_i, the unique part of x_i, is assumed to be uncorrelated with $\xi_1, \xi_2, \ldots, \xi_n$ and with δ_j for $j \neq i$. The unique part δ_i consists of two components: a specific factor s_i and a pure random measurement error e_i. These are indistinguishable, unless the measurements x_i are designed in such a way that they can be separately identified (panel designs and multitrait-multimethod designs). The term δ_i is often called the *measurement error* in x_i even though it is widely recognized that this term may also contain a specific factor as stated above. We shall continue this tradition and use this term in this book.

In a confirmatory factor analysis, the investigator has such knowledge about the factorial nature of the variables that he/she is able to specify that each measure x_i depends only on a few of the factors ξ_j. If x_i does not depend on ξ_j, $\lambda_{ij} = 0$ in (1.3). In a path diagram, this means that there is no one-way arrow from ξ_j to x_i. In many applications, the latent factor ξ_j represents a theoretical construct and the observed measures x_i are designed to be indicators of this construct. In this case there is only one non-zero λ_{ij} in each equation (1.3). This was the case in the previous example.

We shall illustrate confirmatory factor analysis by means of a detailed example. In particular, this example illustrates the assessment of model fit and the use of the model modification index.

Example 5: Nine Psychological Variables - A Confirmatory Factor Analysis

Holzinger & Swineford (1939) collected data on twenty-six psychological tests administered to 145 seventh- and eighth-grade children in the Grant-White school in Chicago. Nine of these tests were selected and for this example it was hypothesized that these measure three common factors: vi-

sual perception (P), verbal ability (V) and speed (S) such that the first three variables measure P, the next three measure V, and the last three measure S. The nine selected variables and their intercorrelations are given in Table 1.5. A path diagram is shown in Figure 1.5.

Table 1.5   Correlation Matrix for Nine Psychological Variables

VIS PERC	1.000								
CUBES	0.318	1.000							
LOZENGES	0.436	0.419	1.000						
PAR COMP	0.335	0.234	0.323	1.000					
SEN COMP	0.304	0.157	0.283	0.722	1.000				
WORDMEAN	0.326	0.195	0.350	0.714	0.685	1.000			
ADDITION	0.116	0.057	0.056	0.203	0.246	0.170	1.000		
COUNTDOT	0.314	0.145	0.229	0.095	0.181	0.113	0.585	1.000	
S-C CAPS	0.489	0.239	0.361	0.309	0.345	0.280	0.408	0.512	1.000

We want to examine the fit of the model implied by the stated hypothesis. If the model does not fit the data well, we want to suggest an alternative model that fits the data better.

The model is very similar to that of the previous example. The only differences are that there are nine observed variables instead of six and that there are three factors instead of two. However, as will be seen, the example will illustrate how the initial model can be evaluated and modified when it is found not to fit the data sufficiently well.

The SIMPLIS input (**EX5A.SPL**) is almost a copy of the input file for Example 4:

```
Nine Psychological Variables - A Confirmatory Factor Analysis

Observed Variables
    'VIS PERC' CUBES LOZENGES 'PAR COMP' 'SEN COMP' WORDMEAN
    ADDITION COUNTDOT 'S-C CAPS'
Correlation Matrix From File EX5.COR
Sample Size 145

Latent Variables: Visual Verbal Speed
```

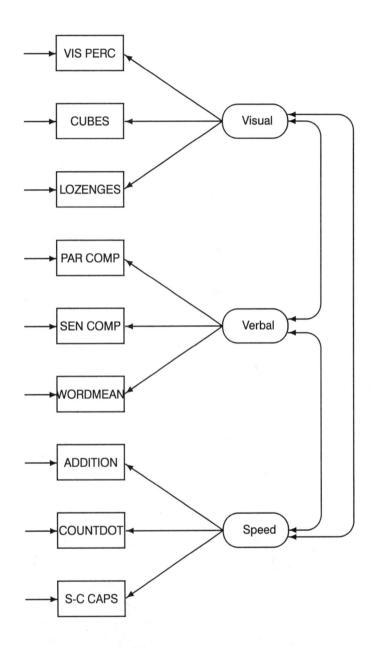

Figure 1.5
Confirmatory Factor Analysis Model for Nine Psychological Variables

```
Relationships:
   'VIS PERC' - LOZENGES = Visual
   'PAR COMP' - WORDMEAN = Verbal
    ADDITION - 'S-C CAPS' = Speed

Number of Decimals = 3
Wide Print
Print Residuals

End of Problem
```

This time we specify the measurement model as relationships rather than as paths, as we did in the previous example. The two new elements in this input file are:

```
Number of Decimals = 3
Wide Print
```

The first of these specifies that we wish to have the results in the output file given by three decimals. Since most users can only interpret at most two decimals, LISREL 8 uses two decimals by default. This example illustrates how this default value can be overridden. The second of these lines is used to specify output with 132 characters per line. Otherwise, there will be 80 characters per line.

Look at the output file obtained for this model. Apart from the estimated relationships and the factor correlation matrix, which all look reasonable, many goodness-of-fit statistics are given. These will be explained in Section 4.5 of Chapter 4. For the moment we will only use the chi-square in the first line:

```
CHI-SQUARE WITH 24 DEGREES OF FREEDOM = 52.626 (P = .000648)
```

A chi-square of 52.63 with 24 degrees of freedom indicates that the model does not fit the data well. How should the model be modified to fit the data better? A very powerful tool for answering this question is the modification index. There is a modification index for each fixed parameter in the model, i.e., for every path that is missing in the path diagram. For each such path, the modification index is an estimate or prediction of the decrease in chi-square that will be obtained if that particular path is introduced in the model. LISREL 8 lists all the large modification indices as follows:

```
        THE MODIFICATION INDICES SUGGEST TO ADD THE
   PATH TO      FROM     DECREASE IN CHI-SQUARE    NEW ESTIMATE
   ADDITION     Visual          10.5                  -.37
   COUNTDOT     Verbal          10.1                  -.28
   S-C CAPS     Visual          24.7                   .57
   S-C CAPS     Verbal          10.0                   .26
```

The largest modification index is 24.7 for the path from Visual to S-C CAPS. This indicates that we can expect a large decrease in chi-square if we include this path in the model. Therefore, *if we can interpret this path substantively* (see below), we can modify the model by adding this path and running the modified model. It is also predicted that the new path will be 0.57.

The fact that the model is misspecified can also be seen from the standardized residuals in the output file:

```
STANDARDIZED RESIDUALS

          VIS PERC   CUBES LOZENGES PAR COMP SEN COMP WORDMEAN ADDITION COUNTDOT S-C CAPS

VIS PERC    .000
   CUBES   -.793    .000
LOZENGES  -1.353   2.087    .000
PAR COMP    .438   -.139    .020    .000
SEN COMP    .020  -1.294   -.565    .416    .000
WORDMEAN    .529   -.609    .950   -.356   -.054    .000
ADDITION  -1.946  -1.762  -3.117    .375   1.225   -.056    .000
COUNTDOT    .884  -1.102  -1.141  -3.013   -.655  -2.084   5.007    .000
S-C CAPS   4.650    .958   2.294   2.241   2.904   1.776  -2.007  -2.886    .000
```

This shows a large standardized residual of 4.65 between VIS PERC and S-C CAPS, indicating that these two variables correlate more than the model accounts for. Although this shows where the lack of fit is, it does not tell how the model should be modified to fit the data better. From this point of view, modification indices are often more useful than standardized residuals for detecting specification errors in the model.

To run the modified model, change the line (see file **EX5B.SPL**)

```
'VIS PERC' - LOZENGES = Visual
```

in the input file to:

```
'VIS PERC' - LOZENGES 'S-C CAPS' = Visual
```

The chi-square for the modified model is:

```
CHI-SQUARE WITH  23 DEGREES OF FREEDOM =    28.862 (P = .185)
```

This indicates that the fit of the modified model is acceptable. Note that the reduction in chi-square is $52.63 - 28.86 = 23.77$ which is roughly the same as what the modification index predicted. Also note that the estimated loading of S-C CAPS on Visual is .46 which is a little smaller than the prediction. The t-value is 5.16. Hence this loading is "significant." Thus, S-C CAPS is not a pure measure of Speed, but rather a composite measure of both Visual and Speed.

The substantive interpretation of the results of these analyses may be as follows. The first factor is "visual perception" as represented by the first three variables containing spatial problems with geometrical configurations. The third factor is a "speed" factor supposed to measure the ability to perform very simple tasks quickly and accurately. However, unlike the two measures "Addition" and "Counting dots," which are purely numerical, variable nine, "Straight-curved capitals" requires the ability to distinguish between capital letters which contain curved parts (like P) from those which contain only straight lines (like L), and to do this quickly and accurately. It is therefore conceivable that "Straight-curved capitals" contains a component correlated with "visual perception" as represented in this data and also that it contains a component of "speed." Thus, the variable "Straight-curved capitals" is a composite measure unlike all the other measures in this example, which are all pure measures.

1.5   Path Analysis with Latent variables

Path analysis with directly observed variables was discussed in Section 1.2. It is also possible to consider path analysis for latent variables. In its most general form there is a structural equation system for a set of latent variables classified as dependent or independent. In most applications, the system is recursive, but models with non-recursive systems have also been proposed. Recursive systems are considered in Examples 6 and 7, and a non-recursive system for latent variables is considered in Example 8.

1.5.1 Recursive System

Recursive models are particularly useful for analyzing data from longitudinal studies in psychology, education, and sociology. In the sociological literature, there have been a number of articles concerned with the specification of models incorporating causation and measurement errors, and analysis of data from panel studies; see Bohrnstedt (1969), Heise (1969, 1970), and Duncan (1969, 1972). Jöreskog & Sörbom (1976, 1977, 1985), Jöreskog (1979b), Jagodzinski & Kühnel (1988), among others, discuss statistical models and methods for analysis of longitudinal data.

The characteristic feature of a longitudinal research design is that the same measurement instruments are used on the same people at two or more occasions. The purpose of a longitudinal or panel study is to assess the changes that occur between the occasions, and to attribute these changes to certain background characteristics and events existing or occurring before the first occasion and/or to various treatments and developments that occur after the first occasion. Often, when the same variables are used repeatedly, there is a tendency for the measurement errors in these variables to correlate over time because of specific factors, memory or other retest effects. Hence there is a need to consider models with correlated measurement errors.

Example 6: Stability of Alienation

Wheaton, et al. (1977) report on a study concerned with the stability over time of attitudes such as alienation, and the relation to background variables such as education and occupation. Data on attitude scales were collected from 932 persons in two rural regions in Illinois at three points in time: 1966, 1967, and 1971. The variables used for the present example are the Anomia subscale and the Powerlessness subscale, taken to be indicators of Alienation. This example uses data from 1967 and 1971 only. The background variables are the respondent's education (years of schooling completed) and Duncan's Socioeconomic Index (SEI). These are taken to be indicators of the respondent's socioeconomic status (Ses). The sample covariance matrix of the six observed variables is given in Table 1.6.

The model to be considered here is shown in Figure 1.6. We specify the error terms of ANOMIA and POWERL to be correlated over time

Table 1.6   Covariance Matrix for Stability of Alienation

	y_1	y_2	y_3	y_4	x_1	x_2
ANOMIA67	11.834					
POWERL67	6.947	9.364				
ANOMIA71	6.819	5.091	12.532			
POWERL71	4.783	5.028	7.495	9.986		
EDUC	−3.839	−3.889	−3.841	−3.625	9.610	
SEI*	−2.190	−1.883	−2.175	−1.878	3.552	4.503

* The variable SEI has been scaled down by a factor 10.

to take specific factors into account. The four one-way arrows on the right side represent the measurement errors in ANOMIA67, POWERL67, ANOMIA71, and POWERL71, respectively. The two-way arrows on the right side indicate that some of these measurement errors are correlated. The covariance between the two error terms for each variable can be interpreted as a specific error variance. For other models for the same data, see Jöreskog & Sörbom (1989, pp. 170–171).

To set up this model for SIMPLIS is straightforward as shown in the following input file (**EX6A.SPL**).

```
Stability of Alienation
Observed Variables
   ANOMIA67   POWERL67   ANOMIA71   POWERL71   EDUC   SEI
Covariance Matrix
   11.834
    6.947      9.364
    6.819      5.091      12.532
    4.783      5.028       7.495      9.986
   -3.839     -3.889      -3.841     -3.625     9.610
   -2.190     -1.883      -2.175     -1.878     3.552    4.503
Sample Size 932
Latent Variables  Alien67 Alien71 Ses
Relationships
   ANOMIA67 POWERL67 = Alien67
   ANOMIA71 POWERL71 = Alien71
   EDUC SEI = Ses
   Alien67 = Ses
   Alien71 = Alien67 Ses
```

```
Let the Errors of ANOMIA67 and ANOMIA71 Correlate
Let the Errors of POWERL67 and POWERL71 Correlate
End of Problem
```

The model is specified in terms of relationships. The first three lines specify the relationships between the observed and the latent variables. The last two lines specify the structural relationships. For example,

```
ANOMIA71 POWERL71 = Alien71
```

means that the observed variables ANOMIA71 and POWERL71 depend on the latent variable Alien71, i.e., that ANOMIA71 and POWERL71 are indicators of Alien71. The line

```
Alien71 = Alien67 Ses
```

means that the latent variable Alien71 depends on the two latent variables Alien67 and Ses. This is one of the two structural relationships.

One can specify the model in terms of its paths instead of its relationships:

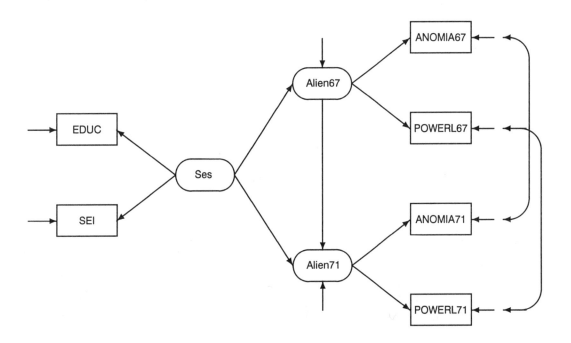

Figure 1.6   Model for Stability of Alienation

```
Paths
   Alien67 -> ANOMIA67 POWERL67
   Alien71 -> ANOMIA71 POWERL71
   Ses -> EDUC SEI
   Alien67 -> Alien71
   Ses -> Alien67 Alien71
```

The output reveals that the model fits very well. Chi-square is 4.73 with 4 degrees of freedom. The two structural equations are estimated as:

```
Alien67 =  - 0.56*Ses, Errorvar.= 0.68, R² = 0.32
           (0.051)
           -11.11


Alien71 = 0.57*Alien67 - 0.21*Ses, Errorvar.= 0.50, R² = 0.50
          (0.055)        (0.046)
          10.27          -4.51
```

The estimated covariance matrix of all the latent variables is also given in the output file:

```
         COVARIANCE MATRIX OF LATENT VARIABLES
            Alien67     Alien71        Ses
Alien67      1.00
Alien71       .68       1.00
    Ses      -.56       -.53       1.00
```

Since the latent variables are standardized in this solution, this is a correlation matrix. The error covariances appear in the output file as

```
Error Covariance for ANOMIA71 and ANOMIA67 = 1.62
                                            (0.31)
                                             5.17


Error Covariance for POWERL71 and POWERL67 = 0.34
                                            (0.26)
                                             1.30
```

The solution just presented is in terms of standardized latent variables. LISREL 8 automatically standardizes all latent variables unless some other units of measurement are specified (see Section 6.9). In this example, when a covariance matrix is analyzed and the units of measurement are the same at the two occasions, it would be more meaningful to assign units of measurement to the latent variables in relation to the observed variables. This will make the two paths from Ses directly comparable.

For this purpose, the relationships should be specified as (see file **EX6B.SPL**):

```
ANOMIA67 = 1*Alien67
POWERL67 = Alien67
ANOMIA71 = 1*Alien71
POWERL71 = Alien71
EDUC = 1*Ses
SEI = Ses

Alien67 = Ses
Alien71 = Alien67 Ses
```

The 1* in the first measurement relation specifies a fixed coefficient of 1 in the relationship between ANOMIA67 and Alien67. The effect of this is to fix the unit of measurement in Alien67 in relation to the unit in the observed variable ANOMIA67. Similarly, in the third relationship, the unit of measurement in Alien71 is fixed in relation to the unit in the observed variable ANOMIA71. Since ANOMIA67 and ANOMIA71 are measured in the same units, this puts Alien67 and Alien71 on the same scale. The fifth relationship specifies Ses to be on the same scale as EDUC.

When the model is estimated with this new set of scales, the results are as follows:

```
Alien67 =  - 0.58*Ses, Errorvar.= 4.85 , R² = 0.32
            (0.056)                (0.47)
            -10.19                 10.35


Alien71 = 0.61*Alien67 - 0.23*Ses, Errorvar.= 4.09 , R² = 0.50
          (0.051)        (0.052)               (0.40)
          11.89          -4.33                 10.10
```

```
           COVARIANCE MATRIX OF LATENT VARIABLES

              Alien67    Alien71      Ses
              --------   --------   --------
   Alien67      7.10
   Alien71      5.20       8.13
       Ses     -3.91      -3.92      6.80
```

It should be emphasized that the two solutions presented here simply represent the same model with the latent variables in different units. The two solutions are equivalent in the sense of goodness-of-fit to the data.

The effect of Ses on Alienation is negative and larger in 1967 than in 1971, as should be expected. The covariance between the measurement errors in POWERL67 and POWERL71 is not significant. Thus, whereas the specific variance in the ANOMIA measure is rather large, there is no evidence of a specific variance in the POWERLESSNESS measure.

Example 7: Performance and Satisfaction

Bagozzi (1980c) formulated a structural equation model to study the relationship between performance and satisfaction in an industrial sales force. His model was designed specifically to answer such questions as: "Is the link between performance and job satisfaction myth or reality? Does performance influence satisfaction, or does satisfaction influence performance?" (Bagozzi, 1980, p. 65). The model we consider here is a modification of Bagozzi's model shown in Figure 1.7. The modified model is based on a reanalysis of Bagozzi's data by Jöreskog & Sörbom (1982), see also Jöreskog & Sörbom (1989, pp. 151–156).

The observed variables are:

- performance measure (PERFORMM)
- job satisfaction measure 1 (JBSATIS1)
- job satisfaction measure 2 (JBSATIS2)
- achievement motivation measure 1 (ACHMOT1)
- achievement motivation measure 2 (ACHMOT2)
- task-specific self-esteem measure 1 (T-S S-E1)
- task-specific self-esteem measure 2 (T-S S-E2)
- verbal intelligence measure (VERBINTM)

The latent variables are:

- performance (Perform)
- job satisfaction (Jobsatis)
- achievement motivation (Achmot)
- task-specific self-esteem (T-s s-e)
- verbal intelligence (Verbint)

Detailed information about the observed variables are given in Bagozzi's article. Table 1.7 gives the means, standard deviations, and product-
moment correlations of the observed variables based on a sample of $N =$
122.

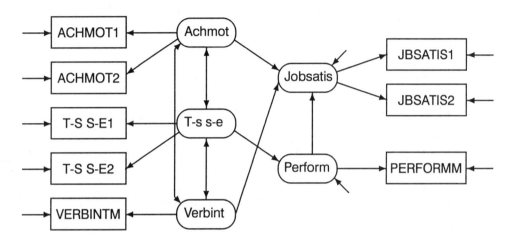

Figure 1.7   Modified Model for Performance and Satisfaction

The model in Figure 1.7 is of the same form as that of Figure 1.6. It
differs in one fundamental aspect—namely that there is only one indicator of the latent variables Verbint and Perform. A consequence of this
is that we cannot estimate the measurement error in the corresponding
observed indicators VERBINTM and PERFORMM. The measurement error variances in these variables are *not* identified parameters. The easiest way to deal with this problem is to assume simply that VERBINTM
equals Verbint and that PERFORMM equals Perform, which is equivalent
to saying that the error variances of VERBINTM and PERFORMM are
zero. The input file is (**EX7A.SPL**):

```
Modified Model for Performance and Satisfaction

Observed Variables: PERFORMM JBSATIS1 JBSATIS2 ACHMOT1 ACHMOT2
                    'T-S S-E1' 'T-S S-E2' VERBINTM
Correlation Matrix from File EX7.DAT
Standard Deviations from File EX7.DAT
Sample Size = 122
Latent Variables: Perform Jobsatis Achmot 'T-s s-e' Verbint
```

Table 1.7
Means, Standard Deviations and Correlations for the Observed Variables in Bagozzi's Model

Variable				*Correlations*				
PERFORMM	1.000							
JBSATIS1	.418	1.000						
JBSATIS2	.394	.627	1.000					
ACHMOT1	.129	.202	.266	1.000				
ACHMOT2	.189	.284	.208	.365	1.000			
T-S S-E1	.544	.281	.324	.201	.161	1.000		
T-S S-E2	.507	.225	.314	.172	.174	.546	1.000	
VERBINTM	−.357	−.156	−.038	−.199	−.277	−.294	−.174	1.000
Means	720.86	15.54	18.46	14.90	14.35	19.57	24.16	21.36
St. Dev.	2.09	3.43	2.81	1.95	2.06	2.16	2.06	3.65

```
Relationships:
    PERFORMM = 1*Perform
    JBSATIS1 = 1*Jobsatis
    JBSATIS2 = Jobsatis
    ACHMOT1  = 1*Achmot
    ACHMOT2  = Achmot
    'T-S S-E1' = 1*'T-s s-e'
    'T-S S-E2' = 'T-s s-e'
    VERBINTM = 1*Verbint

    Perform = 'T-s s-e'
    Jobsatis = Perform Achmot Verbint
Set the Error Variance of PERFORMM to 0
Set the Error Variance of VERBINTM to 0
End of Problem
```

The first line is a title line. The next six lines define the observed and latent variables and read the data and the sample size. The first eight lines of relationships define the measurement equation for each of the observed variables. The 1* are fixed coefficients which define the unit of measurement in the latent variables (see Section 6.9 in Chapter 6). The next two lines define the two structural relationships. The last two lines specify that the error variances of PERFORMM and VERBINTM should be zero.

The data file **EX7.DAT** contains both the correlation matrix and the standard deviations, as follows:

```
1.000
 .418 1.000
 .394  .627 1.000
 .129  .202  .266 1.000
 .189  .284  .208  .365 1.000
 .544  .281  .324  .201  .161 1.000
 .507  .225  .314  .172  .174  .546 1.000
-.357 -.156 -.038 -.199 -.277 -.294 -.174 1.000
2.09 3.43 2.81 1.95 2.06 2.16 2.06 3.65
```

The model fits the data well: Chi-square with 15 degrees of freedom is 14.07. However, one might argue that the verbal intelligence test VERBINTM is a fallible measure and it is therefore unreasonable to assume that its error variance is zero. Instead, we assume that the reliability of VERBINTM is 0.85. It is argued that an arbitrary value of 0.85 is a better assumption than an equally arbitrary value of 1.00. The assumed value of the reliability will affect parameter estimates as well as standard errors. A reliability of 0.85 for VERBINTM is equivalent to an error variance of 0.15 times the variance of VERBINTM. Thus, we assume that the error variance of VERBINTM is $0.15 \times 3.65^2 = 1.998$. In the input file, this is specified as (see file **EX7B.SPL**):

```
Set the Error Variance of VERBINTM to 1.998
```

The two structural equations are estimated as:

```
Perform = 0.92*T-s s-e, Errorvar.= 2.04
         (0.14)                  (0.40)
          6.40                    5.15
```

```
Jobsatis = 0.59*Perform + 1.23*Achmot + 0.21*Verbint, Errorvar.= 3.87
          (0.14)          (0.48)         (0.11)                  (1.22)
           4.24            2.57           2.00                    3.16
```

Our analysis confirms Bagozzi's results with one modification. Although Bagozzi's original model included the hypothesis that verbal intelligence should have a positive effect on performance, he found that this effect was negative and not significant. In our analysis, verbal intelligence has a slightly significant positive effect on satisfaction. The difference in the results is due to our inclusion of 15 percent measurement error in the verbal intelligence measure. The effect of this measurement error in general can be seen in Table 1.8, where we compare the results obtained with and without measurement error in VERBINTM. The effects of measurement error in VERBINTM are rather small.

Table 1.8
Parameter Estimates with and without Measurement Error in VERBINTM

Parameter	*Without measurement error*	*With measurement error*
T-s s-e ⟶ Perform	0.92 (0.14)	0.92 (0.14)
Perform ⟶ Jobsatis	0.58 (0.15)	0.59 (0.14)
Achmot ⟶ Jobsatis	1.18 (0.46)	1.23 (0.48)
Verbint ⟶ Jobsatis	0.18 (0.10)	0.21 (0.11)

1.5.2 Non-Recursive System

The next example appears to be the first non-recursive model in the literature with latent variables. In addition to its non-recursiveness, it involves a feature we have not seen in the previous model: direct effects of observed variables on latent variables. Previously, all paths from latent variables have been directed to observed variables or to other latent variables. But in this example we have both observed "causes" and observed indicators of latent variables.

Example 8: Peer Influences on Ambition

Sociologists have often called attention to the way in which one's peers— e.g., best friends—influence one's decisions—e.g., choice of occupation. They have recognized that the relation must be reciprocal—if my best friend influences my choice, I must influence his. Duncan, Haller, & Portes (1968) present a simultaneous equation model of peer influences on occupational choice, using a sample of Michigan high-school students paired with their best friends. The authors interpret educational and occupational choice as two indicators of a single latent variable "ambition," and specify the choices. This model with reciprocal causation between two latent variables is displayed in Figure 1.8. Note that the variables in this figure are symmetrical with respect to a horizontal line in the middle.

In Figure 1.8 we have omitted all the two-way arrows among the x-variables, i.e., the six variables to the left in the figure. Let

$x_1 =$ respondent's parental aspiration (REPARASP)

$x_2 =$ respondent's intelligence (REINTGCE)

$x_3 =$ respondent's socioeconomic status (RESOCIEC)

$x_4 =$ best friend's socioeconomic status (BFSOCIEC)

$x_5 =$ best friend's intelligence (BFINTGCE)

$x_6 =$ best friend's parental aspiration (BFPARASP)

$y_1 =$ respondent's occupational aspiration (REOCCASP)

$y_2 =$ respondent's educational aspiration (REEDASP)

$y_3 =$ best friend's educational aspiration (BFEDASP)

$y_4 =$ best friend's occupational aspiration (BFOCCASP)

The two latent variables are respondent's ambition (Reambitn) and best friend's ambition (Bfambitn).

The data for this example are given in Table 1.9. Standard deviations are not given in Duncan, Haller, & Portes (1968). Fictitious standard deviations are used here to emphasize that a covariance matrix must be analyzed to obtain correct standard errors of parameter estimates (see Cudeck, 1989).

First, we prepare three files: **EX8.LAB**, **EX8.COR**, and **EX8.STD**. **EX8.LAB** contains the names of the variables in free format, as follows:

```
REINTGCE REPARASP RESOCIEC REOCCASP 'RE EDASP'
BFINTGCE BFPARASP BFSOCIEC BFOCCASP 'BF EDASP'
```

Note that, although long names are used in the path diagram, only eight-character names are used in the file as the program can only use labels of at most eight characters. Note also that the order of the labels in the file corresponds to the order of the variables in the correlation matrix in Table 1.7, not the order of the variables in the path diagram.

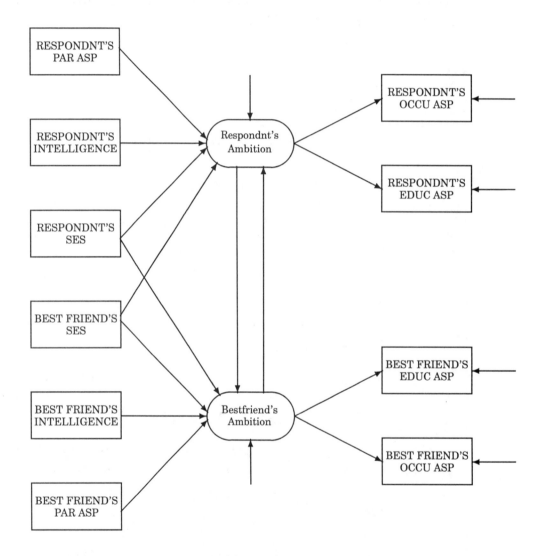

Figure 1.8   Path Diagram for Peer Influences on Ambition

Table 1.9

Correlations and Standard Deviations for Background and Aspiration Measures for 329 Respondents and Their Best Friends

Respondent

REINTGCE	1.0000									
REPARASP	.1839	1.0000								
RESOCIEC	.2220	.0489	1.0000							
REOCCASP	.4105	.2137	.3240	1.0000						
RE EDASP	.4043	.2742	.4047	.6247	1.0000					

Best Friend

BFINTGCE	.3355	.0782	.2302	.2995	.2863	1.0000				
BFPARASP	.1021	.1147	.0931	.0760	.0702	.2087	1.0000			
BFSOCIEC	.1861	.0186	.2707	.2930	.2407	.2950	−.0438	1.0000		
BFOCCASP	.2598	.0839	.2786	.4216	.3275	.5007	.1988	.3607	1.0000	
BF EDASP	.2903	.1124	.3054	.3269	.3669	.5191	.2784	.4105	.6404	1.0000
St. Dev.	1.095	1.071	1.030	.978	1.001	1.036	1.021	1.002	1.055	1.022

The file **EX8.COR** contains the correlation matrix in free format:

```
1.0000
 .1839 1.0000
 .2220  .0489 1.0000
 .4105  .2137  .3240 1.0000
 .4043  .2742  .4047  .6247 1.0000
 .3355  .0782  .2302  .2995  .2863 1.0000
 .1021  .1147  .0931  .0760  .0702  .2087 1.0000
 .1861  .0186  .2707  .2930  .2407  .2950 -.0438 1.0000
 .2598  .0839  .2786  .4216  .3275  .5007  .1988  .3607 1.0000
 .2903  .1124  .3054  .3269  .3669  .5191  .2784  .4105  .6404 1.0000
```

The file **EX8.STD** contains the standard deviations of the variables:

```
1.095 1.071 1.030 .978 1.001 1.036 1.021 1.002 1.055 1.022
```

The input file for the model in Figure 1.8 is as follows (**EX8A.SPL**):

```
Peer Influences on Ambition
- - - - - - - - - - - - - - -
Observed Variables from File EX8.LAB
Correlation Matrix from File EX8.COR
```

```
Standard Deviations from File  EX8.STD
Reorder Variables: REPARASP REINTGCE RESOCIEC BFSOCIEC BFINTGCE
                   BFPARASP REOCCASP 'RE EDASP' 'BF EDASP' BFOCCASP
Sample Size 329
Latent Variables: Reambitn  Bfambitn
Relationships
   REOCCASP  = 1*Reambitn
   'RE EDASP'= Reambitn
   'BF EDASP'= Bfambitn
   BFOCCASP  = 1*Bfambitn
   Reambitn  = Bfambitn REPARASP - BFSOCIEC
   Bfambitn  = Reambitn RESOCIEC - BFPARASP

End of Problem
```

The observed variables may be reordered to correspond with their order in the path diagram. This is done by the two lines:

```
Reorder Variables: REPARASP REINTGCE RESOCIEC BFSOCIEC BFINTGCE
                   BFPARASP REOCCASP 'RE EDASP' 'BF EDASP' BFOCCASP
```

This is not necessary, but it is convenient because it makes the output more easily readable. It will also produce a path diagram which looks similar to that in Figure 1.8, see Chapter 3.

The model is specified in the form of relationships. The first four relationships specify how the y-variables, i.e., the four variables on the right side in the path diagram, depend on the two latent variables. The 1* in the first and the last of these four relationships means that these paths should be fixed at 1.00. This defines the units of measurement of the two latent variables Reambitn and Bfambitn to be the same as that of REOCCASP and BFOCCASP, respectively. Since the observed variables REOCCASP and BFOCCASP are measured in the same unit, this implies that the two latent variables Reambitn and Bfambitn are in the same units also, which is necessary to be able to compare the path coefficients between respondent and best friend.

The last two relationships specify how the two latent variables Reambitn and Bfambitn depend on the x-variables.

In this model we are interested in testing the hypothesis that the effect of Reambitn on Bfambitn is equal to the effect of Bfambitn on Reambitn. This means that we must estimate these two paths under the constraint that they are equal. LISREL handles all such problems under the heading *equality constraints*. For this model, this is achieved by specifying

```
Set Path from Reambitn to Bfambitn equal to Path from Bfambitn to Reambitn
```

or

```
Let Path from Reambitn to Bfambitn = Path from Bfambitn to Reambitn
```

in the input file. The statement "path from A to B" can also be expressed as "path to B from A." There are various alternative ways of specifying the same thing (see Section 6.8 of Chapter 6). Almost any free form may be used as long as:

- ☐ The line begins with Set or Let.

- ☐ There are two pairs of variables mentioned on the line.

The simplest form is perhaps

```
Set Reambitn -> Bfambitn = Bfambitn -> Reambitn
```

where, as before, the symbol -> is used to denote a path.

The model without the equality constraint imposed results in a chi-square of 26.89 with 16 degrees of freedom. When the equality constraint is imposed (see file **EX8B.SPL**), chi-square is 26.96 with 17 degrees of freedom. The difference in chi-square, 0.07, may be used as a chi-square test statistic with one degree of freedom for testing the hypothesis that the reciprocal paths between the latent variables are equal. Thus, there is good evidence that the two reciprocal paths are equal.

One may also be interested in the hypothesis of complete symmetry between best friend and respondent, i.e., that the model is completely symmetric above and below a horizontal line in the middle of the path diagram. To test this hypothesis, one must include all the following equality constraints in the input file (see file **EX8C.SPL**).

```
Set Reambitn -> Bfambitn = Bfambitn -> Reambitn
Set REPARASP -> Reambitn = BFPARASP -> Bfambitn
Set REINTGCE -> Reambitn = BFINTGCE -> Bfambitn
Set RESOCIEC -> Reambitn = BFSOCIEC -> Bfambitn
Set BFSOCIEC -> Reambitn = RESOCIEC -> Bfambitn
Set Reambitn -> 'RE EDASP' = Bfambitn -> 'BF EDASP'
```

The overall chi-square for this model is 32.03 with 22 degrees of freedom. Using the initial model as a base model, the chi-square difference is 5.14 with 6 degrees of freedom, so the hypothesis of complete symmetry cannot be rejected.

This analysis of Model 8C specifies complete symmetry by putting equality constraints on the corresponding paths in the path diagram. It is also reasonable to assume that the error variances are equal between the respondent and the best friend for each observed and latent variable.

To specify equal error variances for the two latent variables, include a line:

```
Set the Error Variances of Reambitn and Bfambitn Equal
```

Similarly, set the error variances equal for each of the observed variables (see file **EX8D.SPL**):

```
Set the Error Variances of REOCCASP and BFOCCASP Equal
Set the Error Variances of 'RE EDASP' and 'BF EDASP' Equal
```

The overall chi-square for Model 8D is 33.44 with 25 degrees of freedom. The difference from the previous model is 1.41 with 3 degrees of freedom. Hence, there is good evidence that the error variances are also equal.

1.6   Analysis of Ordinal variables

In many cases, especially when data are collected through interviews or questionnaires, the observed variables are ordinal, i.e., responses are classified into different ordered categories.

An ordinal variable z (z may be either a y- or an x-variable in LISREL sense) may be regarded as a crude measurement of an underlying unobserved or unobservable continuous variable z^*. For example, a four-point ordinal scale may be conceived as:

- □ If $z^* \leq \tau_1$, z is scored 1
- □ If $\tau_1 < z^* \leq \tau_2$, z is scored 2
- □ If $\tau_2 < z^* \leq \tau_3$, z is scored 3
- □ If $\tau_3 < z^*$, z is scored 4

where $\tau_1 < \tau_2 < \tau_3$ are *threshold* values for z^*. It is often assumed that z^* has a standard normal distribution, in which case the thresholds can be estimated from the inverse of the normal distribution function.

Suppose z_1 and z_2 are two ordinal variables with underlying continuous variables z_1^* and z_2^*, respectively. Assuming that z_1^* and z_2^* have a bivariate normal distribution, their correlation is called the *polychoric correlation coefficient*. A special case of this is the *tetrachoric correlation coefficient* when both z_1 and z_2 are dichotomous. Now, suppose further that z_3 is a continuous variable measured on an interval scale. The correlation between z_1^* and z_3 is called the *polyserial correlation coefficient* assuming that z_1^* and z_3 have a bivariate normal distribution. A special case of this is the *biserial correlation* when z_1 is dichotomous.

An ordinal variable z does not have a metric scale. To use such a variable in a linear relationship we use the corresponding underlying variable z^* instead. The polychoric and polyserial correlations are not correlations computed from actual scores, but are rather theoretical correlations of the underlying z^* variables. These correlations are estimated from the observed pairwise contingency tables of the ordinal variables. See Jöreskog & Sörbom (1996a), Jöreskog (1990), and references therein for the theory on which the polychoric and polyserial correlations are based.

When the observed variables in LISREL are all ordinal or are of mixed scale types (ordinal and interval), the use of ordinary product-moment correlations based on raw scores is not recommended. Instead it is suggested that estimates of polychoric and polyserial correlations be computed and that the matrix of such correlations be analyzed by the WLS method.

The weight matrix required for such an analysis is the inverse of the estimated asymptotic covariance matrix **W** of the polychoric and polyserial correlations. The asymptotic covariance matrix as well as the matrix of polychoric and polyserial correlations are obtained by PRELIS.

The steps involved in this analysis will be described in the following example. This example involves ordinal variables and polychoric correlations only. For other examples involving also continuous and/or censored variables and other types of correlations, see the *PRELIS 2: User's Reference Guide* (Jöreskog & Sörbom, 1996a).

Example 9: Panel Model for Political Efficacy

Aish & Jöreskog (1990) analyze data on political attitudes. Their data consist of 16 ordinal variables measured on the same people at two occasions. Six of the 16 variables were considered to be indicators of Political Efficacy and System Responsiveness. The attitude questions corresponding to these six variables are:

- ❑ *People like me have no say in what the government does (NOSAY)*
- ❑ *Voting is the only way that people like me can have any say about how the government runs things (VOTING)*
- ❑ *Sometimes politics and government seem so complicated that a person like me cannot really understand what is going on (COMPLEX)*
- ❑ *I don't think that public officials care much about what people like me think (NOCARE)*
- ❑ *Generally speaking, those we elect to Parliament lose touch with the people pretty quickly (TOUCH)*
- ❑ *Parties are only interested in people's votes but not in their opinions (INTEREST)*

Permitted responses to these questions were agree strongly, agree, disagree, disagree strongly, don't know, and no answer.

Aish & Jöreskog (1990) considered many different models. Only one of them will be used here. This is shown in Figure 1.9. This includes the idea of correlated measurement errors. Aish & Jöreskog (1990) split the raw data in two random subsamples, an exploration sample of size 410 and a confirmation (validation) sample of size 395.

This model involves two kinds of correlated error terms. The structural residuals are correlated, indicated by the small two-way arrow in the model. The measurement errors in VOTING and COMPLEX are supposed to be correlated over time, indicated by the two long two-way arrows. The errors in VOTING and COMPLEX are said to be *autocorrelated* and the two-way arrows represent the *autocovariance*.

Jöreskog & Sörbom (1996a) illustrate how to use PRELIS to obtain the matrix of polychoric correlations for the exploration sample and how to estimate the asymptotic covariance matrix of the polychoric correlations from the total sample. For our purposes, we assume that the matrix of polychoric correlations, estimated from the exploration sample, has been saved in the file **PANELUSA.PME** and that the corresponding asymptotic covariance matrix, estimated from the total sample, has been saved in the file **PANELUSA.ACP**. The names of the variables are in the file **PANEL.LAB**.

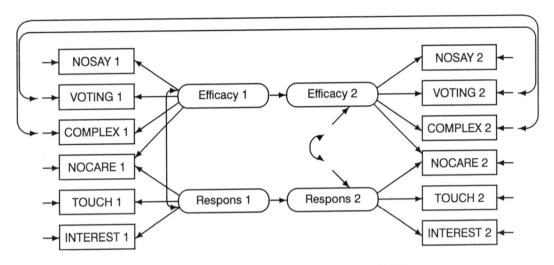

Figure 1.9   Panel Model for Political Efficacy

First, we run the model without correlated measurement errors. The input file is (**EX9A.SPL**):

```
Panel Model for Political Efficacy
See Aish and Joreskog, Quality and Quantity (1990)
Observed Variables from File PANEL.LAB
Correlation Matrix from File PANELUSA.PME
Asymptotic Covariance Matrix from File PANELUSA.ACP
Sample size: 410
Latent Variables: Efficac1 Respons1 Efficac2 Respons2
Relationships:
   NOSAY1 - NOCARE1 = Efficac1
   NOCARE1 - INTERES1 = Respons1

   NOSAY2 - NOCARE2 = Efficac2
   NOCARE2 - INTERES2 = Respons2

   Efficac2 = Efficac1
   Respons2 = Respons1

Let the Errors for Efficac2 and Respons2 correlate

End of Problem
```

The new element here is the line

```
Asymptotic Covariance Matrix from File PANELUSA.ACP
```

The file **PANELUSA.ACP** produced by PRELIS 2 is necessary for the analysis of polychoric correlations. The line above tells LISREL 8 to read the asymptotic covariance matrix from this file and to use WLS (Weighted Least Squares) to estimate the parameters of the model. If no asymptotic covariance matrix is read, LISREL 8 will use ML (maximum likelihood) by default to estimate the parameters, as in all previous examples (see Section 6.14.4 of Chapter 6). Using ML to fit the model to a matrix of polychoric correlations violates the statistical theory on which the estimation of chi-square and standard errors are based.

Since the observed variables are ordinal and have no units of measurement, it is not possible to use any of these as reference variables to assign units of measurement to the latent variables. The most reasonable thing to do is to assume that all variables, observed as well as latent, are standardized.

The first two lines of relationships define the measurement model at time 1, and the next two lines define a similar measurement model at time 2. The following two lines of relationships define the structural relationships among the latent variables. Finally, the last line of relationships specifies that the two structural disturbance terms should be correlated as indicated by the small two-way arrow in the middle of the path diagram.

The output file reveals that the model does not fit the data. Chi-square is 113.83 with 48 degrees of freedom. The P-value for this is 0.00000029. If the model is "true," and all assumptions of the analysis hold, the odds of obtaining such a chi-square are less than 3 in $1,000,000$.

The output file gives the following table of modification indices for error covariances:

```
THE MODIFICATION INDICES SUGGEST TO ADD AN ERROR COVARIANCE
  BETWEEN      AND    DECREASE IN CHI-SQUARE    NEW ESTIMATE
   NOSAY1    VOTING2          13.1                 -.15
  VOTING1    VOTING2          17.9                  .23
  VOTING1     NOSAY1           9.4                  .16
 COMPLEX1   COMPLEX2          43.9                  .36
```

The two largest modification indices are for the covariance of the measurement errors in VOTING1 and VOTING2 and the covariance for the measurement error in COMPLEX1 and COMPLEX2. This suggests that there are large specific factors in VOTING and COMPLEX.

In the next run, we therefore specify the model with the two sets of correlated error terms as shown in Figure 1.9. This is done by adding the two lines

```
Let the errors for VOTING1 and VOTING2 correlate
Let the errors for COMPLEX1 and COMPLEX2 correlate
```

in the input file (see file **EX9B.SPL**).

The chi-square for this model is 50.52 with 46 degrees of freedom which has a $P = .30$. Hence, this model fits the data well. The auto-covariances in VOTING and COMPLEX are given in the output as

```
Error Covariance for VOTING1 and VOTING2 = 0.23
                                          (0.051)
                                           4.54
Error Covariance for COMPLEX1 and COMPLEX2 = 0.35
                                          (0.050)
                                           7.01
```

Both autocovariances are highly significant. Hence, there is strong evidence that there are large specific factors in VOTING and COMPLEX. The results of Model 9B are summarized in Tables 1.10–1.13.

Table 1.10   Loadings and their Standard Errors

	Time 1		*Time 2*	
	Efficacy	*Respons*	*Efficacy*	*Respons*
NOSAY	.85 (.05)		.79 (.08)	
VOTING	.58 (.06)		.71 (.08)	
COMPLEX	.56 (.05)		.47 (.06)	
NOCARE	.37 (.11)	.50 (.11)	.42 (.12)	.49 (.12)
TOUCH		.80 (.04)		.85 (.05)
INTEREST		.92 (.03)		.89 (.06)

Table 1.11   Error Variances, Reliabilities and Autocovariances

	Error variances		Reliabilities		Autocovariances
	Time 1	Time 2	Time 1	Time 2	Specific Variances
NOSAY	.28	.37	.72	.63	
VOTING	.43	.26	.57	.74	.23(.05)
COMPLEX	.34	.43	.66	.57	.35(.05)
NOCARE	.32	.27	.68	.73	
TOUCH	.36	.29	.64	.71	
INTEREST	.15	.20	.85	.80	

Table 1.12   Correlations between Efficacy and Responsiveness

Time 1	Time 2
.78	.81

Table 1.13   Stability Coefficients and Residual Covariance Matrix

	Stability Coefficients	Residual Covariance Matrix	
Efficacy	.72(.09)	.48	
Respons	.64(.07)	.45	.59

2 MULTI-SAMPLE EXAMPLES

In the previous chapter we have shown how simple LISREL models may be specified and estimated using the SIMPLIS command language. Thus far, all of the examples have been based on data from a single sample. But the SIMPLIS language can also be used to analyze data from several samples simultaneously, according to a multiple-group LISREL model with some or all parameters constrained to be equal over groups. Examples of such simultaneous analyses have been given by Jöreskog (1971), McGaw & Jöreskog (1971), Sörbom (1974, 1975, 1976, 1978, 1981), Sörbom & Jöreskog (1981), Werts, Rock, Linn, & Jöreskog (1976, 1977), Alwin & Jackson (1981), Mare & Mason (1981), and Lomax (1983) among others.

Consider a set of G populations. These may be different nations, states or regions, culturally or socioeconomically different groups, groups of individuals selected on the basis of some known or unknown selection variables, groups receiving different treatments, and control groups, etc. In fact, they may be any set of mutually exclusive groups of individuals that are clearly defined. It is assumed that a number of variables have been measured on a number of individuals from each population. This approach is particularly useful in comparing a number of treatment and control groups regardless of whether individuals have been assigned to the groups randomly or not.

Any LISREL model may be specified and fitted for each group of data. However, LISREL 8 assumes by default that the models are identically the same over groups, i.e., all relationships and all parameters are the same in each group. Thus, only differences between groups need to be specified. Our first example of a multi-group analysis clarifies how this is done.

2.1 Equal Factor Structures

Example 10: Testing Equality of Factor Structures

Table 2.1 gives observed covariance matrices for two samples ($N_1 = 865$, $N_2 = 900$, respectively) of candidates who took the Scholastic Aptitude Test in January 1971. The four measures are, in order, VERBAL40 = a 40-item verbal aptitude section, VERBAL50 = a separately timed 50-item verbal aptitude section, MATH35 = a 35-item math aptitude section, and MATH25 = a separately timed 25-item math aptitude section.

Werts, Rock, Linn, & Jöreskog (1977) used these data to test various assumptions about the psychometric properties of the tests both within and between groups. Werts, Rock, Linn, & Jöreskog (1976) showed how to compare correlations, variances, covariances and regression coefficients between groups. Here we use the data to illustrate how one can test equality of factor structures in a confirmatory factor analysis model.

We regard VERBAL40 and VERBAL50 as indicators of a latent variable Verbal and MATH35 and MATH25 as indicators of a latent variable

Table 2.1   Covariance Matrices for SAT Verbal and Math Sections

Covariance Matrix for Group 1

Tests	VERBAL40	VERBAL50	MATH35	MATH25
VERBAL40	63.382			
VERBAL50	70.984	110.237		
MATH35	41.710	52.747	60.584	
MATH25	30.218	37.489	36.392	32.295

Covariance Matrix for Group 2

Tests	VERBAL40	VERBAL50	MATH35	MATH25
VERBAL40	67.898			
VERBAL50	72.301	107.330		
MATH35	40.549	55.347	63.203	
MATH25	28.976	38.896	39.261	35.403

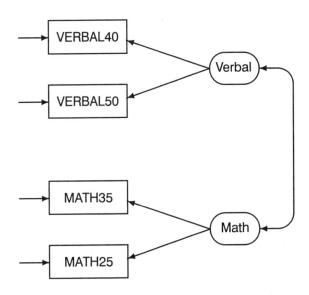

Figure 2.1   Path Diagram for SAT Verbal and Math

Math. The model we consider is shown in Figure 2.1. The model is a simple confirmatory factor analysis model of the same type as in Example 4.

There are three sets of parameters in the model: (1) the four factor loadings corresponding to the paths from Verbal and Math to the observed variables, (2) the correlation between Verbal and Math, and (3) the four error variances of the observed variables. We want to investigate to what extent each of these sets of parameters are invariant over groups.

We begin by constructing a data file **EX10.COV** containing the two covariance matrices. In free format, each covariance matrix can be written on one line. File **EX10.COV** looks like this:

```
63.382 70.984 110.237 41.710 52.747 60.584 30.218 37.489 36.392 32.295
67.898 72.301 107.330 40.549 55.347 63.203 28.976 38.896 39.261 35.403
```

In the first model (Model A), we assume that all parameters are the same in both groups. The input file **EX10A.SPL** is:

```
Group 1: Testing Equality Of Factor Structures
Model A: Factor Loadings, Factor Correlation, Error Variances Invariant
Observed Variables: VERBAL40 VERBAL50 MATH35 MATH25
Covariance Matrix from File EX10.COV
```

```
Sample Size = 865
Latent Variables: Verbal Math
Relationships:
   VERBAL40 VERBAL50 = Verbal
   MATH35 MATH25 = Math

Group 2: Testing Equality Of Factor Structures
Covariance Matrix from File EX10.COV
Sample Size = 900
End of Problem
```

The new element here is that we are reading specifications for two groups in the same input file. For each group, the input is written as described in the previous chapter. To specify the beginning of each group, the first word in the first title line for each group begins with the word Group. The specifications for groups 2, 3, 4, ..., may be considerably simplified, because of the general principle that everything is the same as in the previous group unless otherwise stated. Thus, in the input above, only two things differ between group 1 and group 2—namely, the data in the covariance matrices and the sample sizes. The names of the variables, observed as well as latent, are the same, and the model is the same. Notice that no relationships are specified for group 2, which implies that they are the same as for group 1.

The output file reveals two solutions, one for each group, but the parameter estimates are all identical. The value of chi-square is reported only after the second group. In a multi-sample analysis, the chi-square is a measure of fit of all models in all groups, and, in general, this chi-square cannot be decomposed into a chi-square for each group separately.

For our example, chi-square is 34.89 with 11 degrees of freedom, so we must seek a less constrained model. Suppose therefore that we allow the factor loadings to be different for the two groups, retaining the invariance of the factor correlation and the error variances. The input for such a model is (**EX10B.SPL**):

```
Group 1: Testing Equality Of Factor Structures
Model B : Factor Correlation and Error Variances Invariant

Observed Variables: VERBAL40 VERBAL50 MATH35 MATH25
Covariance Matrix from File EX10.COV
Sample Size = 865

Latent Variables: Verbal Math
```

```
Relationships:
   VERBAL40 VERBAL50 = Verbal
   MATH35 MATH25 = Math

Group 2: Testing Equality Of Factor Structures
Covariance Matrix from File EX10.COV
Sample Size = 900
Relationships:
   VERBAL40 VERBAL50 = Verbal
   MATH35 MATH25 = Math
End of Problem
```

The only difference between this and the previous input is that the relationships of the model are also specified in the second group. Note that this implies that the coefficients in the relationships will be estimated for the second group independently of the first. But since nothing is specified about the factor correlation and the error variances, these are assumed to be the same for both groups.

The chi-square for Model B is 29.67 with seven degrees of freedom, which also does not indicate a good fit. Continuing the example, let us assume that, in addition to the factor loadings, the error variances may also be different over groups, still retaining the assumption that the factor correlation is the same. To do so, add the line

```
Set the Error Variances of VERBAL40 - MATH25 free
```

before the last line, see **File EX10C.SPL**. This allows the error variances for the second group to be different from those of the first group.

Model C gives a chi-square of 4.03 with three degrees of freedom ($P = 0.26$), which indicates a good fit. The difference in chi-square between Models B and C is 25.64 with four degrees of freedom, and that between Models A and B is 5.23 with four degrees of freedom. This suggests that it is really the error variances that differ between groups and not the factor loadings. We may therefore want to do one final analysis of this data, namely one in which factor loadings and factor correlations are the same for both groups and only the error variances are allowed to be different. This may be done with the following input file (**EX10D.SPL**):

```
Group 1: Testing Equality Of Factor Structures
Model D: Factor Loadings and Factor Correlation Invariant
```

```
Observed Variables: VERBAL40 VERBAL50 MATH35 MATH25
Covariance Matrix from File EX10.COV
Sample Size = 865
Latent Variables: Verbal Math
Relationships:
   VERBAL40 VERBAL50 = Verbal
   MATH35 MATH25 = Math

Group 2: Testing Equality Of Factor Structures
Covariance Matrix from File EX10.COV
Sample Size = 900
Set the Error Variances of VERBAL40 - MATH25 free
End of Problem
```

The chi-square for this model is 10.87 with seven degrees of freedom ($P = 0.14$). Note that this model fits the data, whereas Model B—with the same degrees of freedom—does not.

Example 11: Parental Socioeconomic Characteristics

An interesting example of multi-sample analyses is given by Mare & Mason (1981) in a study of the reliabilities of son's and parents' reports on father's and mother's education and occupation. They report the covariance matrices in Table 2.2 for six variables and three populations. The sample size is 80 for each group. The variables are:

SOFED = Son's report of father's education

SOMED = Son's report of mother's education

SOFOC = Son's report of father's occupation

FAFED = Father's report of his own education

MOMED = Mother's report of her own education

FAFOC = Father's report of his own occupation

The three populations are:

Group 1: White Sixth Graders

Group 2: White Ninth Graders

Group 3: White Twelfth Graders

Table 2.2
Covariance Matrices for Parental Socioeconomic Characteristics

	1.	2.	3.	4.	5.	6.
Sixth Grade						
1. SOFED	5.86					
2. SOMED	3.12	3.32				
3. SOFOC	35.28	23.85	622.09			
4. FAFED	4.02	2.14	29.42	5.33		
5. MOMED	2.99	2.55	19.20	3.17	4.64	
6. FAFOC	35.30	26.91	465.62	31.22	23.38	546.01
Ninth Grade						
1. SOFED	8.20					
2. SOMED	3.47	4.36				
3. SOFOC	45.65	22.58	611.63			
4. FAFED	6.39	3.16	44.62	7.32		
5. MOMED	3.22	3.77	23.47	3.33	4.02	
6. FAFOC	45.58	22.01	548.00	40.99	21.43	585.14
Twelfth Grade						
1. SOFED	5.74					
2. SOMED	1.35	2.49				
3. SOFOC	39.24	12.73	535.30			
4. FAFED	4.94	1.65	37.36	5.39		
5. MOMED	1.67	2.32	15.71	1.85	3.06	
6. FAFOC	40.11	12.94	496.86	38.09	14.91	538.76

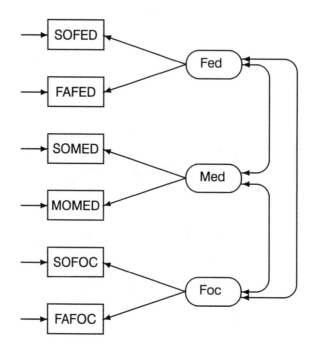

Figure 2.2   Path Diagram for Parental Socioeconomic Characteristics

The model is shown in Figure 2.2, where the latent variables Fed, Med, and Foc represent the true father's education, mother's education, and father's occupation, respectively.

It seems reasonable to assume that the covariance matrix of the latent variables Fed, Med, and Foc are the same in all groups and that the error variances of FAFED, MOMED, and FAFOC should be the same in all groups but that the error variances of SOFED, SOMED, and SOFOC should vary over groups. Such a model may be specified as follows (**EX11A.SPL**):

```
Group 1: Reports of Parental Socioeconomic Characteristics - Grade 6
Observed Variables: SOFED SOMED SOFOC FAFED MOMED FAFOC
Covariance Matrix
5.86 3.12 3.32 35.28 23.85 622.09 4.02 2.14 29.42 5.33
2.99 2.55 19.20 3.17 4.64 35.30 26.91 465.62 31.22 23.38 546.01
Sample Size: 80
Latent Variables: Fed Med Foc
SOFED = Fed
SOMED = Med
SOFOC = Foc
FAFED = 1*Fed
MOMED = 1*Med
FAFOC = 1*Foc


Group 2: Reports of Parental Socioeconomic Characteristics - Grade 9
Covariance Matrix
8.20 3.47 4.36 45.65 22.58 611.63 6.39 3.16 44.62 7.32
3.22 3.77 23.47 3.33 4.02 45.58 22.01 548.00 40.99 21.43 585.14
SOFED = Fed
SOMED = Med
SOFOC = Foc
Let the Error Variances of SOFED - SOFOC be free


Group 3: Reports of Parental Socioeconomic Characteristics - Grade 12
Covariance Matrix
5.74 1.35 2.49 39.24 12.73 535.30 4.94 1.65 37.36 5.39
1.67 2.32 15.71 1.85 3.06 40.11 12.94 496.86 38.09 14.91 538.76
SOFED = Fed
SOMED = Med
SOFOC = Foc
Let the Error Variances of SOFED - SOFOC be free
End of Problem
```

Note the following:

□ Since the sample size is the same for all groups, it need only be specified for the first group.

□ The units of measurements of the latent variables Fed, Med, and Foc are defined in the first group by using FAFED, MOMED, and FAFOC as reference variables (see Section 6.9 of Chapter 6). Since this is not repeated in the second and third group, it stays the same. This defines the units of the latent variables to be the same in all groups, which is essential. Otherwise, it would not make sense to postulate equal variances and covariances for the latent variables.

□ Since nothing is specified about the variances and covariances of the latent variables, their covariance matrix will be the same in all groups.

□ the inclusion of the line

```
Let the Error Variances of SOFED - SOFOC be free
```

for the second and third group makes the error variances of SOFED, SOMED, and SOFOC free in all groups.

Chi-square for this model is 78.04 with 36 degrees of freedom. Mare & Mason (1981) considered another model in which the error terms of SOMED and SOFED were allowed to correlate in the first two groups but not in the third. To specify this, add the line

```
Set the Error Covariance between SOMED and SOFED free
```

in the first two groups, and add the line

```
Set the Error Covariance between SOMED and SOFED equal to 0
```

in the third group, see **EX11B.SPL**. The last line is needed; otherwise, the error covariance between SOMED and SOFED in the third group would be estimated to be equal to that of the second group.

Chi-square for this model is 52.73 with 34 degrees of freedom. The output file gives the reliabilities (squared multiple correlations) shown in Table 2.3. This is a somewhat remarkable result because in grade 12, the sons' reports of their fathers' education and occupation are more reliable than the fathers' reports of their own education and occupation. In the earlier ages, however, the sons' reports are less reliable than their parents' reports, as expected.

Table 2.3

Estimated Reliabilities of Son's and Parents' Reports of Parental Socioeconomic Characteristics

	SOFED	SOMED	SOFOC	FAFED	MOMED	FAFOC
Grade 6	.62	.38	.71	.87	.93	.90
Grade 9	.76	.86	.92	.87	.93	.90
Grade 12	.91	.81	.95	.87	.93	.90

2.2 Equal Regressions

In this section, we consider the problem of testing whether a regression equation is the same in several populations. Suppose that a dependent variable y and a number of explanatory variables $x_1, x_2, \ldots, x_q$ are observed in two or more groups. We are interested in determining the extent to which the regression equation

$$y = \alpha + \gamma_1 x_1 + \gamma_2 x_2 + \cdots + \gamma_q x_q + z \qquad (2.1)$$

is the same in different groups. We say that the regressions are:

- ☐ equal if $\alpha, \gamma_1, \gamma_2, \ldots, \gamma_q$, are the same in all groups
- ☐ parallel if $\gamma_1, \gamma_2, \ldots, \gamma_q$, are the same in all groups

Normally the covariance matrix of the x-variables is not expected to be the same across groups and often one finds that the intercept terms differ between groups. It may also be the case that *only some* of the regression coefficients are the same across groups.

Example 12: Testing Equality of Regressions

Sörbom (1976) gave the covariance matrices in Table 2.4. These are based on scores on the ETS Sequential Test of Educational Progress (STEP) for two groups of boys who took the test in both Grade 5 and Grade 7. The two groups were defined according to whether or not they were in the academic curriculum in Grade 12. We will use this data to demonstrate how one can test the equality of various regressions.

Table 2.4
Means and Covariance Matrices for STEP Reading and Writing

Boys Academic ($N = 373$)

STEP Reading, Grade 5	281.349			
STEP Writing, Grade 5	184.219	182.821		
STEP Reading, Grade 7	216.739	171.699	283.289	
STEP Writing, Grade 7	198.376	153.201	208.837	246.069
Means	262.236	258.788	275.630	269.075

Boys Non-Academic ($N = 249$)

STEP Reading, Grade 5	174.485			
STEP Writing, Grade 5	134.468	161.869		
STEP Reading, Grade 7	129.840	118.836	228.449	
STEP Writing, Grade 7	102.194	97.767	136.058	180.460
Means	248.675	246.896	258.546	253.349

We begin by typing the data and saving it for both groups in file
EX12.DAT in the form of covariance matrices and means. It should look
like this:

```
281.349
184.219 182.821
216.739 171.699 283.289
198.376 153.201 208.837 246.069
262.236 258.788 275.630 269.075
174.485
134.468 161.869
129.840 118.836 228.449
102.194  97.767 136.058 180.460
248.675 246.896 258.546 253.349
```

As a first example, we consider the testing of equal regressions of STEP
Reading at Grade 7 on STEP Reading and Writing at Grade 5. STEP
Writing at Grade 7 is not involved in this example. For example, we may
want to predict STEP Reading at Grade 7 from STEP Reading and Writing

at Grade 5 and to see if the prediction equation is the same for both groups. A path diagram is shown in Figure 2.3, where CONST is a variable which is constant equal to 1 for every case. The intercept α in the regression equation is the coefficient of CONST.

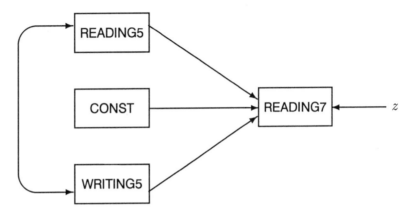

Figure 2.3   Path Diagram for Regression of READING7

The input file for this example is (**EX12A.SPL**):

```
Group BA: STEP Reading and Writing, Grades 5 and 7
Observed Variables: READING5 WRITING5 READING7 WRITING7
Covariance Matrix from File EX12.DAT
Means from File EX12.DAT
Sample Size: 373
Equation: READING7 = CONST READING5 WRITING5

Group BNA: STEP Reading and Writing, Grades 5 and 7
Covariance Matrix from File EX12.DAT
Means from File EX12.DAT
Sample Size: 249
Set the Error Variance of READING7 free
End of Problem
```

Note the following:

❑ Four variables are read but only three variables are used in the model. The program automatically selects the variables for analysis on the basis of the variables included on the Equation line (see Section 6.4.6).

❑ The intercept α in the regression equation is the regression coefficient of the variable CONST. The line

```
Equation: READING7 = CONST READING5 WRITING5
```

is interpreted in the usual way, i.e., three coefficients α, γ_1, and γ_2 will be estimated. In order to estimate the intercept α, the means of the observed variables must be provided. If only covariance matrices are given, α will be zero.

❑ The line

```
Set the Error Variance of READING7 Free
```

is needed in the second group; otherwise the regression error variance will be assumed to be the same in both groups.

The test of equal regressions gives a chi-square of 35.73 with three degrees of freedom. To test for parallel regressions rather than equal regressions, one must allow α to be different in the two groups. This is done by adding the line

```
Set the Path from CONST to READING7 Free
```

in the second group (see file **EX12B.SPL**). Another way of specifying the same thing is to include the relationship

```
READING7 = CONST
```

in the second group. The meaning of this is that α, the coefficient of CONST, will be reestimated in the second group, while the remaining part of the regression equation remains the same as in group 1, i.e., the coefficients of READING5 and WRITING5 will be the same in both groups.

The test of parallel regressions gives a chi-square of 3.31 with two degrees of freedom. Thus, it is evident that the regressions are parallel but not equal.

As a second example, consider estimating two regression equations simultaneously in two groups and testing whether both regressions are parallel in the two groups. With the same data as in the previous example, we may estimate the regressions of READING7 and WRITING7 on READING5 and WRITING5 (**EX12C.SPL**):

```
Group BA: STEP Reading and Writing, Grades 5 and 7
Observed Variables: READING5 WRITING5 READING7 WRITING7
Covariance Matrix from File EX12.DAT
Means from File EX12.DAT
Sample Size: 373
Equations: READING7 - WRITING7 = CONST READING5 WRITING5
Set the Error Covariance between READING7 and WRITING7 free

Group BNA: STEP Reading and Writing, Grades 5 and 7
Covariance Matrix from File EX12.DAT
Means from File EX12.DAT
Sample Size: 249
Equations: READING7 - WRITING7 = CONST
Set the Error Variances of READING7 - WRITING7 Free
Set the Error Covariance between READING7 and WRITING7 Free
End of Problem
```

☐ The line

```
Set the Error Covariance between READING7 and WRITING7 Free
```

in the first group, specifies that the two error terms z_1 and z_2 may be correlated, i.e., we don't believe that READING5 and WRITING5 will account for the whole correlation between READING7 and WRITING7.

☐ The line

```
Equations: READING7 - WRITING7 = CONST
```

in the second group, specifies that the intercept terms for the second group do not have to equal those of the first group.

☐ The line

```
Set the Error variances of READING7 - WRITING7 Free
```

in the second group specifies that the variances of z_1 and z_2 in the second group are not constrained to be equal to those of the first group.

☐ The line

```
Set the Error Covariance between READING7 and WRITING7 Free
```

in the second group specifies that the covariance between z_1 and z_2 in the second group is not constrained to be equal to that of the first group.

Chi-square for this model is 9.01 with four degrees of freedom and a
P-value of 0.06.

Another way of modeling the data in Table 2.3 is to take measurement
error in the observed variables into account and to regard Reading and
Writing as two indicators of a latent variable Verbal Ability, say, and to
estimate the regression of Verbal Ability at Grade 7 on Verbal Ability at
Grade 5. This will be considered as Example 15 in Section 2.4. However,
first we must learn how to estimate means of latent variables.

2.3   Estimation of Means of Latent Variables

Since a latent variable is unobservable, it does not have an intrinsic scale.
Neither the origin nor the unit of measurement are defined. In a single
population the origin is fixed by assuming that all observed variables are
measured in deviations from their means and that the means of all latent
variables are zero. The unit of measurement of each latent variable is
usually fixed either by assuming that it is a standardized variable with
variance 1 or by fixing a non-zero loading for a reference variable.

In multi-group studies, these restrictions can be relaxed somewhat by
assuming that the latent variables are on the same scale in all groups.
The common scale may be defined by assuming that the means of the latent variables are zero in one group and that the loadings of the observed
variables on the latent variables are invariant over groups, with one loading for each latent variable fixed for a reference variable. Under these
assumptions it is possible to estimate the means and covariance matrices
of the latent variables relative to this common scale. To illustrate this
we will first consider a small example based on the same data used in the
previous section, and then consider a large example.

Example 13: Mean Difference in Verbal Ability

Using the variables and data in Table 2.4, take READING5 and WRIT-
ING5 to be indicators of a latent variable Verbal5 (Verbal Ability at Grade 5)
and estimate the mean difference in Verbal5 between groups.

The measurement model for READING5 and WRITING5 is:

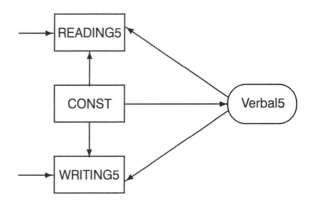

Figure 2.4   Path Diagram for Estimating Mean of Verbal5

$$\text{READING5} = \nu_1 + \lambda_1 \text{Verbal5} + D1$$

$$\text{WRITING5} = \nu_2 + \lambda_2 \text{Verbal5} + D2$$

Note that these relationships now have intercept terms ν_1 and ν_2. The scale for Verbal5 is fixed by assuming that the mean is zero in Group BA and that $\lambda_1 = 1$. It is further assumed that $\nu_1, \nu_2, \lambda_1, \lambda_2$ are invariant over groups. We can then estimate the mean of Verbal5 in Group BNA as well as all the other parameters. The mean of Verbal5 is interpreted as the mean difference in verbal ability between the groups.

A path diagram is shown in Figure 2.4. Note that the path from CONST to READING5 corresponds to ν_1, the path from CONST to WRITING5 corresponds to ν_2, and the path from CONST to Verbal5 corresponds to the mean of Verbal5.

The input file for this model is (**EX13A.SPL**):

```
Group BA: STEP Reading and Writing, Grades 5 and 7
Observed Variables: READING5 WRITING5 READING7 WRITING7
Covariance Matrix from File EX12.DAT
Means from File EX12.DAT
Sample Size: 373
Latent Variable: Verbal5
Relationships:
   READING5 = CONST + 1*Verbal5
   WRITING5 = CONST + (1)*Verbal5
```

```
Group BNA: STEP Reading and Writing, Grades 5 and 7
Covariance Matrix from File EX12.DAT
Means from File EX12.DAT
Sample Size: 249
Relationship: Verbal5 = CONST
Set the Error Variances of READING5 - WRITING5 free
Set the Variance of Verbal5 free
End of Problem
```

The two measurement equations correspond to:

```
READING5 = CONST + 1*Verbal5
WRITING5 = CONST + (1)*Verbal5
```

The intercept terms ν_1 and ν_2 are the coefficients of CONST in these relationships. In the first relationship, λ_1 is given as a fixed coefficient of 1. In the second relationship, the 1 in parenthesis is provided as a starting value for λ_2.

The mean of Verbal5 is the coefficient of CONST in the expression

```
Relationship: Verbal5 = CONST
```

in the second group. Since this is not present in the first group, the mean of Verbal5 will be zero in the first group but estimated in the second group. The variance of Verbal5 is estimated in each group.

The output file shows that the mean of Verbal5 is -13.56 with a standard error of 1.21, from which we can conclude that the non-academic group is below the academic group in verbal ability as measured by these two tests.

As a second example of estimating means of latent variables, we extend the previous model to include also READING7 and WRITING7 as indicators of Verbal7. A path diagram is shown in Figure 2.5, where, for simplicity, we have omitted the variable CONST. It is tacitly assumed that there is a path from CONST to all the variables in the path diagram (as in Figure 2.4). We now estimate the mean difference between groups in two variables, Verbal5 and Verbal7.

The input file for this analysis is (**EX13B.SPL**):

```
Group BA: STEP Reading and Writing, Grades 5 and 7
Observed Variables: READING5 WRITING5 READING7 WRITING7
Covariance Matrix from File EX12.DAT
Means from File EX12.DAT
Sample Size: 373
```

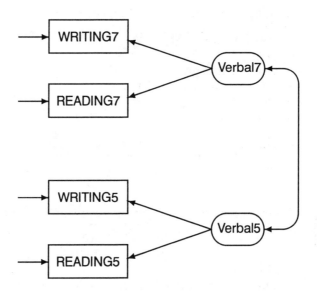

Figure 2.5
Path Diagram for Estimating the Mean of Verbal5 and Verbal7

```
Latent Variables: Verbal5 Verbal7
Relationships:
   READING5 = CONST + 1*Verbal5
   WRITING5 = CONST + (1)*Verbal5
   READING7 = CONST + 1*Verbal7
   WRITING7 = CONST + (1)*Verbal7

Group BNA: STEP Reading and Writing, Grades 5 and 7
Covariance Matrix from File EX12.DAT
Means from File EX12.DAT
Sample Size: 249
Relationships:
   Verbal5 = CONST
   Verbal7 = CONST
Set the Error Variances of READING5 - WRITING7 free
Set the Variances of Verbal5 - Verbal7 free
Set the Covariance of Verbal5 and Verbal7 free
End of Problem
```

The estimated group means, variances and covariances of the two latent variables are shown in Table 2.5. It can be seen that the non-academic group mean is below the mean of the academic group both in Grade 5 and in Grade 7. Furthermore, the non-academic group has a less favorable development from Grade 5 to Grade 7; its mean has decreased compared to the academic group.

Table 2.5
Estimated Means and Covariance Matrices of Verbal5 and Verbal7

Boys Academic (N = 373)		
	Verbal5	Verbal7
Verbal5	220.06	
Verbal7	212.11	233.59
Means	0.00	0.00

Boys Non-Academic (N = 249)		
	Verbal5	Verbal7
Verbal5	156.34	
Verbal7	126.96	153.73
Means	−13.80	−17.31

One way to describe the group differences graphically is to draw 95 percent confidence regions as shown in Figure 2.6. These confidence regions are ellipses, such that 95 percent of the population is located within the ellipse. The ellipses can differ in origin, shape and orientation. For any score on Verbal5, one can see the most likely range of values of Verbal7 for each group.

It can be seen from Figure 2.6 that the slope of the regression lines of Verbal7 on Verbal5 is similar. This leads to the problem of estimating these regression lines and testing whether they are the same, a topic that will be considered in Example 15 of Section 2.4.

As a third example of estimating factor means, we consider a large example involving nine observed variables, three latent variables and four groups. Altogether, this example involves 57 parameters to be estimated.

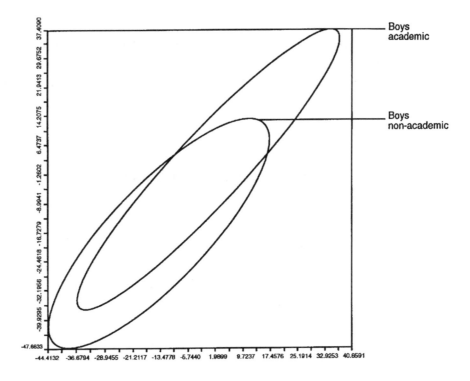

Figure 2.6   Confidence Ellipses for Verbal5 and Verbal7

Example 14: Nine Psychological Variables with Factor Means

Sörbom (1974) used nine selected variables from the classical study of Holzinger & Swineford (1939) to illustrate his methodology. The nine variables were selected to measure three latent variables: Space, Verbal, and Memory. The groups consist of eighth-grade children from two schools in Chicago: the Pasteur and the Grant-White schools. The children from each school were divided into two groups according to whether they scored above or below the median on a speeded addition test. Thus the groups are:

1. *Pasteur Low ($N_1 = 77$)*
2. *Pasteur High ($N_2 = 79$)*
3. *Grant-White Low ($N_3 = 74$)*
4. *Grant-White High ($N_4 = 71$)*

The variables, correlations, standard deviations, and means are given in Tables 2.6 and 2.7. The standard deviations have been scaled so that a weighted average of the within group covariance matrices is a correlation matrix.

Table 2.6   Nine Psychological Variables: Correlations

Group 1 above main diagonal; Group 2 below main diagonal

	1	2	3	4	5	6	7	8	9
Visual Perception	—	.32	.48	.28	.26	.40	.42	.12	.23
Cubes	.24	—	.33	.01	.01	.26	.32	.05	−.04
Paper Form Board	.23	.22	—	.06	.01	.10	.22	.03	.01
General Information	.32	.05	.23	—	.75	.60	.15	−.08	−.05
Sentence Completion	.35	.23	.18	.68	—	.63	.07	.06	.10
Word Classification	.36	.10	.11	.59	.66	—	.36	.19	.24
Figure Recognition	.22	.01	−.07	.09	.11	.12	—	.29	.19
Object-Number	−.02	−.01	−.13	.05	.08	.03	.19	—	.38
Number-Figure	.09	−.14	−.06	.16	.02	.12	.15	.29	—

Group 3 above main diagonal; Group 4 below main diagonal

	1	2	3	4	5	6	7	8	9
Visual Perception	—	.34	.41	.38	.40	.42	.35	.16	.35
Cubes	.32	—	.21	.32	.16	.13	.27	.01	.27
Paper Form Board	.34	.18	—	.31	.24	.35	.30	.09	.09
General Information	.31	.24	.31	—	.69	.55	.17	.31	.34
Sentence Completion	.22	.16	.29	.62	—	.65	.20	.30	.27
Word Classification	.27	.20	.32	.57	.61	—	.31	.34	.27
Figure Recognition	.48	.31	.32	.18	.20	.29	—	.31	.38
Object-Number	.20	.01	.15	.06	.19	.15	.36	—	.38
Number-Figure	.42	.28	.40	.11	.07	.18	.35	.44	—

The nine labels are stored in the file **EX14.LAB**. Note that the labels must be enclosed within single quotes if they contain blank spaces.

Table 2.7
Nine Psychological Variables: Means and Standard Deviations

	Standard Deviations				*Means*			
	1	2	3	4	1	2	3	4
Visual Perception	1.06	0.96	0.95	1.03	4.20	4.30	4.29	4.21
Cubes	1.20	0.86	1.03	0.86	5.25	5.03	5.32	5.33
Paper Form Board	1.02	0.99	0.92	1.06	4.96	5.06	5.02	5.09
General Information	1.03	0.96	0.99	1.01	2.98	3.41	3.72	4.15
Sentence Completion	1.08	1.06	0.96	0.91	3.20	3.38	3.78	3.88
Word Classification	0.99	1.01	0.95	1.05	4.45	4.76	5.17	5.59
Figure Recognition	1.17	1.01	0.81	0.98	13.42	13.62	13.70	13.72
Object-Number	1.00	1.10	0.83	1.04	1.74	2.14	1.30	1.78
Number-Figure	1.04	1.00	0.88	1.07	2.10	2.16	1.87	2.44

```
'VIS PERC' CUBES 'PAP FORM' 'GEN INFO' 'SENT COM' WORDCLAS 'FIG REC'
'OBJ NUM' 'NUM FIG'
```

All the remaining data are stored in the file **EX14.DAT** in the order: correlations for group 1, standard deviations for group 1, means for group 1, correlations for group 2, standard deviations for group 2, etc. In this data file, we use FORTRAN formats to record the data. Each group of data begins with a format line. After the last right parenthesis in the format, one can write any text, e.g., to identify the particular part of the data for which the format is intended. After the format the data follows. The file **EX14.DAT** looks like this:

```
(21F3.2/24F3.2) Group 1: Pasteur Low: Correlations
100 32100 48 33100 28 01 06100 26 01 01 75100 40 26 10 60 63100
 42 32 22 15 07 36100 12 05 03-08 06 19 29100 23-04 01-05 10 24 19 38100
(9F3.2) Group 1: Pasteur Low: Standard deviations
106120102103108 99117100104
(9F4.2) Group 1: Pasteur Low: Means
 420 525 496 298 320 4451342 174 210
(21F3.2/24F3.2) Group 2: Pasteur High: Correlations
100 24100 23 22100 32 05 23100 35 23 18 68100 36 10 11 59 66100
 22 01-07 09 11 12100-02-01-13 05 08 03 19100 09-14-06 16 02 12 15 29100
(9F3.2) Group 2: Pasteur High: Standard deviations
```

```
 96 86 99 96106101101110100
(9F4.2) Group 2: Pasteur High: Means
 430 503 506 341 338 4761362 214 216
(21F3.2/24F3.2) Group 3: Grant-White Low: Correlations
100 34100 41 21100 38 32 31100 40 16 24 69100 42 13 35 55 65100
 35 27 30 17 20 31100 16 01 09 31 30 34 31100 35 27 09 34 27 27 38 38100
(9F3.2) Group 3: Grant-White Low: Standard deviations
 95103 92 99 96 95 81 83 88
(9F4.2) Group 3: Grant-White Low: Means
 429 532 502 372 378 5171370 130 187
(21F3.2/24F3.2) Group 4: Grant-White High: Correlations
100 32100 34 18100 31 24 31100 22 16 29 62100 27 20 32 57 61100
 48 31 32 18 20 29100 20 01 15 06 19 15 36100 42 28 40 11 07 18 35 44100
(9F3.2) Group 4: Grant-White High: Standard deviations
103 86106101 91105 98104107
(9F4.2) Group 4: Grant-White High: Means
 421 533 509 415 388 5591372 178 244
```

The model is similar to Example 5 of Chapter 1 (see Figure 1.5), but in addition to factor loadings, factor variances and covariances, and error variances, we are now also estimating intercept terms in the measurement relationships and means of the factors. The intercept terms, the loadings, and the error variances are assumed to be invariant over groups. The observed variables VIS PERC, GEN INFO, and FIG REC are used as reference variables for the latent variables. Consequently their loadings are fixed at 1. The means of the latent variables are assumed to be zero in group 1 (Pasteur Low). The input file is (**EX14.SPL**):

```
Group Pasteur Low
Observed Variables from File EX14.LAB
Sample size 77
Covariance Matrix from File EX14.DAT
Standard Deviations from File EX14.DAT
Means from File EX14.DAT
Latent Variables: Space Verbal Memory

Relationships:
    'VIS PERC' - 'PAP FORM' = CONST Space
    'GEN INFO' - WORDCLAS= CONST Verbal
    'FIG REC' - 'NUM FIG' = CONST Memory
    'VIS PERC' = 1*Space
    'GEN INFO' = 1*Verbal
    'FIG REC' = 1*Memory
```

```
Group Pasteur High
Sample Size 79
Covariance Matrix from File EX14.DAT
Standard Deviations from File EX14.DAT
Means from File EX14.DAT
Relationships:
    Space Verbal Memory = CONST
Set the Variances of Space - Memory Free
Set the Covariances of Space - Memory Free

Group Grant-White Low
Sample Size 74
Covariance Matrix from File EX14.DAT
Standard Deviations from File EX14.DAT
means from File EX14.DAT
Relationships:
    Space Verbal Memory = CONST
Set the Variances of Space - Memory Free
Set the Covariances of Space - Memory Free

Group Grant-White High
Sample Size 71
Covariance Matrix from File EX14.DAT
Standard Deviations from File EX14.DAT
Means from File EX14.DAT
Relationships:
    Space Verbal Memory    = CONST
Set the Variances of Space - Memory Free
Set the Covariances of Space - Memory Free
End of Problem
```

The maximum likelihood solution is shown in Table 2.8. Scaled factor means have been computed such that the weighted mean (weighted by sample size) over the groups is zero for each factor. Mean profiles are given in Figure 2.7.

It is seen that the profiles are similar within schools, but with those scoring high on the addition test appearing at a higher level. Spatial ability does not differentiate the groups. For verbal ability, there is a difference between schools, with students from Grant-White being superior. This reflects the fact that the Pasteur school "enrolls children of factory workers, a large percentage of whom were foreign-born and the Grant-White school enrolls children in a middle-class suburban area" (Meredith, 1964). With regard to the memory factor, both groups of the Pasteur

Table 2.8
Maximum Likelihood Estimates for Nine Psychological Variables with Factor Means

Solution Standardized to Common Metric
Factor Loadings

Test	Space	Verbal	Memory	Error Variance	Intercept
Visual Perception	1.00	0	0	0.48	4.20
Cubes	0.58	0	0	0.82	5.20
Paper Form Board	0.71	0	0	0.73	5.00
General Information	0	1.00	0	0.34	3.05
Sentence Completion	0	0.94	0	0.34	3.08
Word Classification	0	0.95	0	0.43	4.50
Figure Recognition	0	0	1.00	0.78	13.55
Object–Number	0	0	1.18	0.76	1.68
Number–Figure	0	0	1.31	0.63	2.06

Factor Covariance Matrices

School	Low Level			High Level		
	0.71			0.37		
Pasteur	0.26	0.76		0.30	0.70	
	0.20	0.09	0.25	0.01	0.08	0.20
	0.46			0.53		
Grant-White	0.38	0.63		0.32	0.63	
	0.19	0.20	0.13	0.33	0.13	0.29

	Factor Means			*Scaled Factor Means*		
	Space	Verbal	Memory	Space	Verbal	Memory
Pasteur Low	0.00	0.00	0.00	−0.05	−0.49	−0.06
Pasteur High	0.05	0.32	0.17	0.00	−0.17	0.12
Grant-White Low	0.09	0.70	−0.14	0.04	0.21	−0.20
Grant-White High	0.06	1.02	0.20	0.01	0.52	0.13

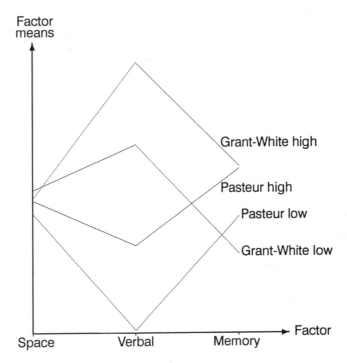

Figure 2.7   Factor Mean Profiles

school seem superior to the groups of the Grant-White school, although the difference between the high groups is small.

2.4   Regression Models with Latent Variables

Example 15: Regression of Verbal7 on Verbal5

Using the variables and data in Table 2.4, take READING5 and WRITING5 to be indicators of a latent variable Verbal5 (Verbal Ability at Grade 5), READING7 and WRITING7 to be indicators of a latent variable Verbal7 (Verbal Ability at Grade 7) and estimate the regression of Verbal7 on Verbal5 for both groups.

The measurement model is the same as in Example 13 (see Figure 2.5). The regression of Verbal7 on Verbal5 is of the form:

$$\texttt{Verbal7} = \alpha + \gamma \texttt{Verbal5} + z \qquad (2.2)$$

Since the means of Verbal7 and Verbal5 are only determined relative to a fixed origin for both groups, it follows that we can set $\alpha = 0$ in the first group and estimate α in the second group. If the slope γ is the same in both groups, we can interpret α as an effect due to group differences.

The input file for this analysis is (**EX15A.SPL**):

```
Group BA: STEP Reading and Writing, Grades 5 and 7
Observed Variables: READING5 WRITING5 READING7 WRITING7
Covariance Matrix from File EX12.DAT
Means from File EX12.DAT
Sample Size: 373
Latent Variables: Verbal5 Verbal7
Relationships:
    READING5 = CONST + 1*Verbal5
    WRITING5 = CONST + (1)*Verbal5
    READING7 = CONST + 1*Verbal7
    WRITING7 = CONST + (1)*Verbal7
    Verbal7 = Verbal5

Group BNA: STEP Reading and Writing, Grades 5 and 7
Covariance Matrix from File EX12.DAT
Means from File EX12.DAT
Sample Size: 249
Relationships:
    Verbal5 = CONST
    Verbal7 = CONST + Verbal5
Set the Error Variances of READING5 - WRITING7 free
Set the Variance of Verbal5 free
Set the Error Variance of Verbal7 free
End of Problem
```

Here, the line Verbal7 = Verbal5 in the first group corresponds to (2.2) with $\alpha = 0$ for the first group. The line Verbal7 = CONST + Verbal5 in the second group corresponds to (2.2) with a different γ estimated in the second group. The line Verbal5 = CONST in the second group will make the program estimate the mean of Verbal5 in the second group as in the previous examples. This mean is zero in group 1.

The two lines

```
Set the Variance of Verbal5 free
Set the Error Variance of Verbal7 free
```

in the second group indicate that we are not assuming the variance of Verbal5 to be the same in the two groups, neither are we assuming that the regression residual variance is the same.

The chi-square for this model is 10.11 with six degrees of freedom ($P = 0.12$). This chi-square is the same as for Example 13B. In fact, these two models are just two different parametrizations of the means, variances, and covariances in Table 2.5. The two slopes are estimated as $\hat{\gamma}^{(1)} = 0.96(0.04)$ and $\hat{\gamma}^{(2)} = 0.81(0.06)$. Approximate 95 percent confidence intervals for the slopes would be $0.88 \leq \gamma^{(1)} \leq 1.04$ and $0.69 \leq \gamma^{(2)} \leq 0.93$. Since these overlap to some extent, it is of interest to test the hypothesis $\gamma^{(1)} = \gamma^{(2)}$.

To do so, replace the line Verbal7 = CONST + Verbal5 in the second group with Verbal7 = CONST (see file **EX15B.SPL**).

This means that α will be estimated as the coefficient of CONST as before. But since there is a path from Verbal5 to Verbal7 in group 1, this path will remain invariant over groups. In this way, the slope γ is estimated to be the same in both groups. Had we included Verbal5 as a second variable on the right side, we would have estimated a different γ in the second group, as we did before.

Chi-square for this model is 15.36 with seven degrees of freedom. The difference in chi-square between this model and the previous model is 5.24 with one *df*. Thus, the hypothesis of equal slopes cannot be rejected at the 1% level. The estimate of α is -4.60 with a *t*-value of -5.12.

Example 16: Head Start Summer Program

Sörbom (1981) reanalyzed some data from the Head Start summer program previously reanalyzed by Magidson (1977). Sörbom used data on 303 white children consisting of a Head Start sample ($N = 148$) and a matched Control sample ($N = 155$). The correlations, standard deviations and means are given in Table 2.9. The children were matched on gender and kindergarten attendance but no attempt had been made to match on social status variables. The variables used in Sörbom's reanalysis were:

$x_1 = $ *Mother's education*

$x_2 = $ *Father's education*

$x_3 = $ *Father's occupation*

$x_4 = $ *Family income*

$y_1 = $ *Score on the Metropolitan Readiness Test*

$y_2 = $ *Score on the Illinois Test of Psycholinguistic Abilities*

Table 2.9
Correlations, Standard Deviations and Means for the Head Start Data

Control Group

Variable	Correlations						St. Dev.	Means
x_1	1.000						1.360	3.839
x_2	.484	1.000					1.195	3.290
x_3	.224	.342	1.000				1.193	2.600
x_4	.268	.215	.387	1.000			3.239	6.435
y_1	.230	.215	.196	.115	1.000		3.900	20.415
y_2	.265	.297	.234	.162	.635	1.000	2.719	10.070

Head Start Group

Variable	Correlations						St. Dev.	Means
x_1	1.000						1.332	3.520
x_2	.441	1.000					1.281	3.081
x_3	.220	.203	1.000				1.075	2.088
x_4	.304	.182	.377	1.000			2.648	5.358
y_1	.274	.265	.208	.084	1.000		3.764	19.672
y_2	.270	.122	.251	.198	.664	1.000	2.677	9.562

We want to do the following:

A   *Test whether x_1, x_2, x_3, and x_4 can be regarded as indicators of a single construct Ses (socioeconomic status) for both groups. Is the measurement model the same for both groups? Is there a difference in the mean of Ses between groups?*

B   *Assuming that y_1 and y_2 can be used as indicators of another construct Ability (cognitive ability), test whether the same measurement model applies to both groups. Test the hypothesis of no difference in the mean of Ability between groups.*

C   *Estimate the structural equation:*

$$\text{Ability} = \alpha + \gamma\text{Ses} + z$$

Is γ the same for the two groups? Test the hypothesis $\alpha = 0$. Interpret the results.

The data file (**EX16.DAT**) is:

```
1
 .484 1
 .224 .342 1
 .268 .215 .387 1
 .230 .215 .196 .115 1
 .265 .297 .234 .162 .635 1
1.360   1.195   1.193   3.239   3.900   2.719
3.839   3.290   2.600   6.435  20.415  10.070

1
 .441 1
 .220 .203 1
 .304 .182 .377 1
 .274 .265 .208 .084 1
 .270 .122 .251 .198 .664 1
1.332   1.281   1.075   2.648   3.764   2.677
3.520   3.081   2.088   5.358  19.672   9.562
```

The model for Problem A is similar to that of Example 13. We have four indicators of a single latent variable Ses and we want to estimate the mean difference in Ses. The input file is (**EX16A.SPL**):

```
Group = Control. Head Start Summer Program - Problem A
Observed Variables: MOTHEDUC FATHEDUC FATHOCCU FAMILINC MRT ITPA
Correlations from File EX16.DAT
Standard Deviations from File EX16.DAT
Means from File EX16.DAT
Sample Size: 149
Latent Variable: Ses
Relationships:
    MOTHEDUC            = CONST + 1*Ses
    FATHEDUC - FAMILINC = CONST +   Ses

Group = Head Start. Head Start Summer Program - Problem A
Correlations from File EX16.DAT
Standard Deviations from File EX16.DAT
Means from File EX16.DAT
Sample Size: 156
Relationship:
    Ses = CONST
```

```
Set the Error Variances of MOTHEDUC - FAMILINC free
Set the Variance of Ses free
End of Problem
```

Note that six variables are read but only four variables are used in the model. The program automatically selects the variables for analysis on the basis of the variables included in the relationships, see Section 6.4.6.

For Problem A we obtain an overall chi-square measure of goodness-of-fit of the model equal to 35.80 with ten degrees of freedom, indicating that the fit of the model is not very good. An examination of the modification indices reveals that the errors of MOTHEDUC and FATHEDUC may be correlated. This indicates that parents' education levels correlate more than can be explained by social status. Thus, to add the covariance between the errors of MOTHEDUC and FATHEDUC, we add the line

```
Let the Errors of MOTHEDUC FATHEDUC correlate
```

in both groups (see **File EX16B.SPL**). This model has an acceptable fit: chi-square with eight degrees of freedom equals 6.46. The difference in degrees of freedom from the previous model is two, as we have added two parameters. The output shows that the groups differ significantly in Ses; the difference is -0.34 with a standard error equal to 0.10. The children of the Head Start Group have a lower Ses on average than the children of the Control Group.

As criteria, Magidson (1977), used two cognitive ability tests—the Metropolitan Readiness Test (MRT) and the Illinois Test of Psycholinguistic Abilities (ITPA). Magidson made separate analyses for the two tests, but here, as suggested by Bentler & Woodward (1978), we will use the two tests to define the construct cognitive ability. This model is the same as in Problem A, except that there are only two observed variables. We leave it to the reader to set up the input file. The model has no degrees of freedom, so one could compute the estimates simply by equating the first and second order moments implied by the model to their observed counterparts. Also, in cognitive ability the Head Start group is below the Control group in the sense that the estimated mean difference is negative (-0.74). However, the difference is not significant; the standard error of the estimate equals 0.44.

For Problem C we use the combined model as depicted in Figure 2.8. The main focus is on the structural equation

$$\texttt{Ability} = \alpha^{(g)} + \gamma^{(g)}\texttt{Ses} + z \ ,$$

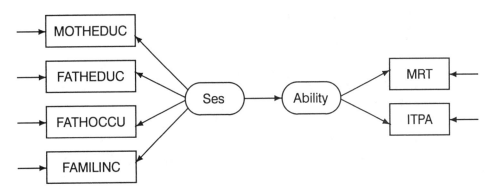

Figure 2.8   Head Start: Model for Problem C

where $g = 1$ for the Control Group and $g = 2$ for the Head Start Group. We set $\alpha^{(1)} = 0$ and estimate $\alpha^{(2)}$, $\gamma^{(1)}$, and $\gamma^{(2)}$.

The input file is (**EX16C.SPL**):

```
Group = Control. Head Start Summer Program - Problem C
Observed Variables: MOTHEDUC FATHEDUC FATHOCCU FAMILINC MRT ITPA
Correlations from File EX16.DAT
Standard Deviations from File EX16.DAT
Means from File EX16.DAT

Sample Size: 149

Latent Variables: Ses Ability

Relationships:
    MOTHEDUC              = CONST + 1*Ses
    FATHEDUC - FAMILINC = CONST +   Ses
    MRT                   = CONST + 1*Ability
    ITPA                  = CONST +   Ability
    Ability = Ses

Let the Errors of MOTHEDUC and FATHEDUC correlate

Group = Head Start. Head Start Summer Program - Problem C
Correlations from File EX16.DAT
Standard Deviations from File EX16.DAT
Means from File EX16.DAT

Sample Size: 156
```

```
Relationships:
    Ses    = CONST
    Ability = CONST + Ses

Set the Error Variances of MOTHEDUC - ITPA free
Set the Variance of Ses free
Set the Error Variance of Ability free
Let the Errors of MOTHEDUC and FATHEDUC correlate
End of Problem
```

The chi-square for the combined model equals 27.87 with 22 degrees of freedom ($P = 0.18$), so the fit of the model is acceptable. Examination of the γ parameters in the two groups shows that they are probably equal, since $\hat{\gamma}^{(1)} = 2.02$ and $\hat{\gamma}^{(2)} = 2.29$ with estimated standard errors equal to 0.63 and 0.73, respectively. Thus, the final model is a model with the γ's constrained to be equal. To specify this, we replace the line Ability = CONST + Ses with the line Ability = CONST (see file **EX16D.SPL**).

The chi-square for this model is 27.99 with 23 degrees of freedom ($P = 0.22$). Note that the P-value increased despite the reduction of parameters. The difference in chi-square for the last two models can be used as a test of the hypothesis that the γ's are equal. Chi-square with one degree of freedom is 0.11. Thus we can treat the γ's as equal. As the regression lines are parallel, it is meaningful to talk about $\alpha^{(2)}$ as a measure of the effect of Head Start.

There seems to be no significant effect of the Head Start program when controlling for social status, although the inclusion of social class has changed the negative effect to be positive. The estimate of $\alpha^{(2)}$ is 0.19 with a standard error equal to 0.38.

In the more general case, when there are more than two groups and/or more than one dependent variable, one can test the hypothesis of no effect by reestimation of the model after adding the restriction $\alpha^{(g)} = 0$, and then compare the chi-squares. In the above case, we obtain chi-square equal to 28.23 with 24 degrees of freedom. Thus, the test of no effect results in a chi-square with one degree of freedom equal to 0.24, which in this case is the same as the square of the t-value of α.

3 PATH DIAGRAMS

In the examples of the two previous chapters we have used path diagrams to present the basic conceptual ideas of a model. From these path diagrams the relationships of the model can easily be written in the input file.

Path diagrams can also be obtained as output from LISREL 8. These features may not be available on all systems, and the appearance of the path diagrams and the way they are produced may vary slightly between systems. We describe them here as they function in the DOS version of LISREL 8.

Several different path diagrams can be produced on the screen, and these path diagrams can be changed interactively to define a modified model, which in turn can be estimated directly without changing the input file. The path diagrams and how to change them are described and illustrated in this chapter using some of the examples from Chapters 1 and 2. Whereas the previous chapters can be read without actually running LISREL 8, this chapter assumes that the reader is running the program on a computer and following the instructions and hints given. Either upper or lower case letters may be entered.

To produce a path diagram of an estimated model, include a line

`Path Diagram`

in the input file. This line may be inserted anywhere after the line

`Observed Variables`

and before the line

`End of Problem`

In the examples that follows, we have inserted the line immediately before the line

`End of Problem`

3.1 Parameter Estimates and *t*-Values

Consider Example 1 (Regression of GNP) and input file **EX1A.SPL**. Insert
the line

```
Path Diagram
```

and run the problem. The path diagram depicted in Figure 3.1 appears
on the screen. This is the path diagram with parameter estimates.

Press **T** (for **T**-values). A similar path diagram appears but with *t*-
values instead of parameter estimates (see Figure 3.2). The non-signi-
ficant *t*-values, if any, will appear in a different color, see Section 3.12.1.

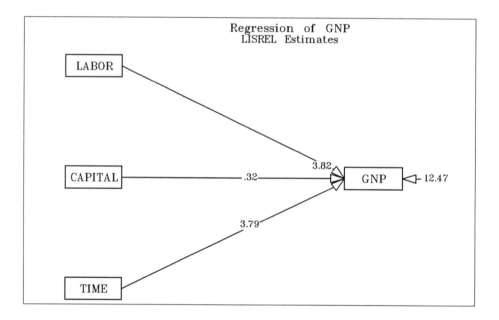

Figure 3.1   Diagram with Parameter Estimates for EX1A.SPL

Press **E** (for Parameter **E**stimates), to get back to the estimates. Press
Q (for **Q**uit) to quit looking at the path diagram.

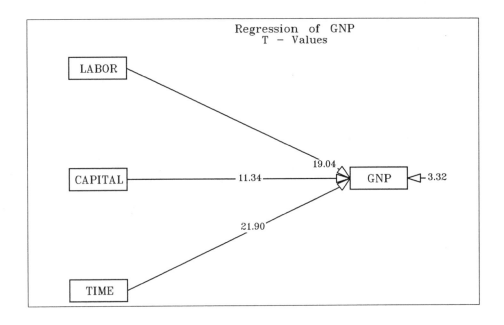

Figure 3.2   Diagram with *t*-values for EX1A.SPL

3.2   B-, X-, Y-, S-, and R-Diagrams

Consider Example 6 (Stability of Alienation) and input file **EX6B.SPL**. Insert the line

```
Path Diagram
```

and run the problem. A path diagram appears on the screen (see Figure 3.3). This is the **B**-diagram with parameter estimates (**B** for **B**asic Model). Error terms are not included in this path diagram.

There are four additional path diagrams with estimates that are produced, but only one is visible at any time. You will see each of the five path diagrams by pressing **S** (for **S**tructural Relationships), **X** (for **X**-Measurement Relationships), **Y** (for **Y**-Measurement Relationships), **R** (for E**R**ror Covariances), and **B** (for **B**asic Model). These will now be explained.

The latent variables are classified into dependent and independent latent variables. The dependent latent variables are those which depend on other latent variables. In the path diagram these have one or more

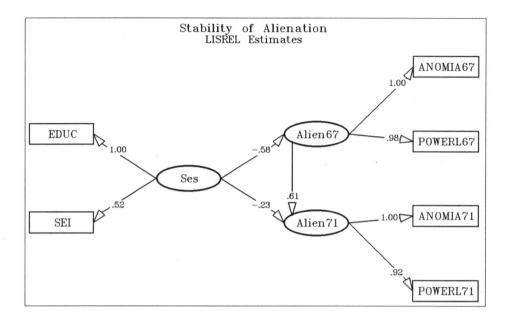

Figure 3.3   B-diagram with Estimates for EX6B.SPL

one-way (unidirected) arrows pointing towards them. Hence the dependent latent variables are Alien67 and Alien71 in this example. The independent latent variables are those which do not depend on other latent variables. In the path diagram they have no one-way arrows pointing to them. In this case there is one independent latent variable, namely Ses. In standard LISREL terminology, the dependent latent variables are called η-variables or ETA-variables, and the independent latent variables are called ξ-variables or KSI-variables.

Next, the observed variables are classified into two categories: y-variables and x-variables. The y-variables are those observed variables that depend on the dependent latent variables, and the x-variables are those that depend on the independent latent variables. In this case, the y-variables are ANOMIA67 ANOMIA71 POWERL67 POWERL71, as these depend on the dependent latent variables Alien67 and Alien71; the x-variables are EDUC and SEI, as these depend on the independent latent variable Ses.

Back to the path diagram, press **S** (for **S**tructural Relationships). The *structural equation model* for the latent variables appears (see Figure 3.4).

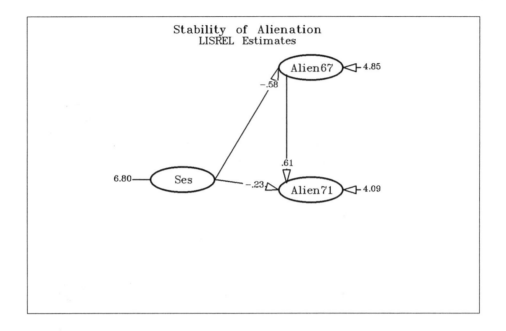

Figure 3.4   S-diagram with Estimates for EX6B.SPL

This is the **S**-diagram which describes how the dependent latent variables depend on the independent latent variables. Note that there is an error term on each of the dependent latent variables and the variances of these appear in the path diagram. Also note that the variance of the latent independent variable appears at a line (not an arrow) attached to Ses. If there are several independent latent variables, all their variances and covariances will be indicated in the path diagram.

Press **X** (for **X**-Measurement Relationships), and the *measurement model for the x-variables* appears (see Figure 3.5).

This is the **X**-diagram which describes how the x-variables depend on the independent latent variables. The measurement errors of the x-variables and their variances are included in this path diagram. The variance of Ses is also shown in this path diagram.

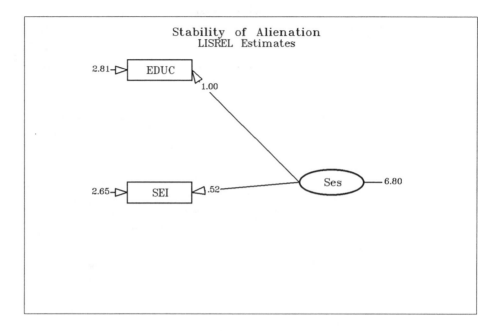

Figure 3.5   X-diagram with Estimates for EX6B.SPL

Press **Y** (for **Y**-Measurement Relationships), and the *measurement model for the y-variables* appears (see Figure 3.6). This is the **Y**-diagram which describes how the y-variables depend on the dependent latent variables. The measurement errors of the y-variables and their variances are included in this path diagram as well as the variances and covariances of the latent dependent variables.

Although the measurement errors of the y-variables are shown in the **Y**-diagram, this diagram does not show that some of these errors are correlated, as was specified in the input file **EX6B.SPL**. To see the correlated error terms, press **R** (for E**R**ror Covariances). This gives a path diagram of the form shown in Figure 3.7. This is the **R**-diagram. Here a two-way arrow indicates that two error terms are correlated and the number attached to the line is the covariance between the two error terms. Note that it is the covariance of the error terms of the variables that are indicated in the path diagram, not the covariance between the observed variables.

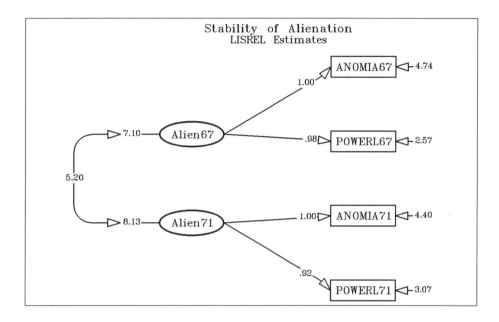

Figure 3.6   Y-diagram with Estimates for EX6B.SPL

The variances and covariances of the measurement errors of the y-variables can be seen in the **R**-diagram. For example, the measurement error variance of ANOMIA67 is 4.74 and that of ANOMIA71 is 4.40. The covariance of these two error terms is also seen in the **R**-diagram and is given as 1.62.

Press **B** for (Basic Model). Then the basic, full model appears again. This combines all three model parts **Y**, **X**, and **S** into one single, simplified path diagram. One can switch between any one of the five path diagrams, by pressing the keys **S**, **X**, **Y**, **R**, **B** and for each such path diagram one can see the parameter estimates or the t-values by pressing **E** (for **E**stimates) or **T** (for **T**-values), respectively. t-values will be given for all estimated parameters except for the variances and covariances of the latent dependent variables.

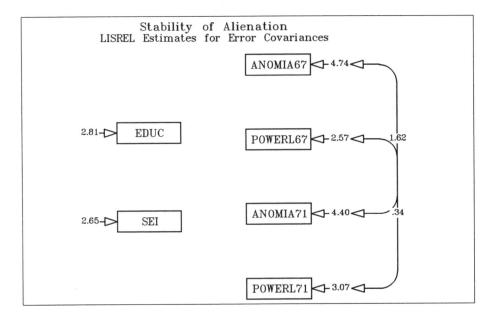

Figure 3.7   R-diagram with Estimates for EX6B.SPL

3.3 Fit Statistics

Consider Example 5 (Nine Psychological Variables - A Confirmatory Factor Analysis) and input file **EX5A.SPL**. Insert the line   .

```
Path Diagram
```

and run the problem. The basic path diagram will appear on the screen. From Section 1.4 we know that the chi-square measure of overall fit for this model is 52.626 with 24 degrees of freedom. Suppose we want to check this and still retain the path diagram we have on the screen. This can be done by pressing **F** (for **F**it Measures) or **G** (for **G**oodness-of-Fit Measures), which gives a long list of fit measures. These will be explained in Chapter 4. For the moment, it will be sufficient to note the value of chi-square which is the first fit measure listed. Pressing any key brings back the previous path diagram. Thus, at any path diagram, one may ask for the fit measures to be displayed and then go back to the path diagram.

3.4   Modification Indices

We continue the example of the previous section. With the basic path diagram on the screen, press **M** (for **M**odification Indices) and a path diagram with modification indices will appear. This is an **M**-diagram which looks like the picture shown in Figure 3.8.

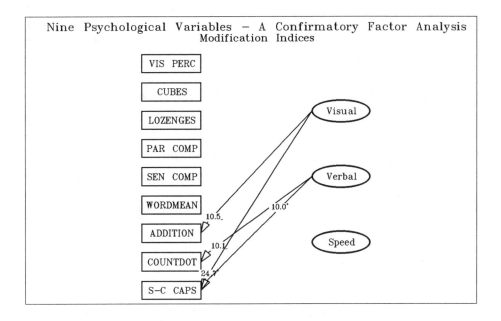

Figure 3.8   B-diagram with Modification Indices for EX5A.SPL

Recall that there is a modification index for each fixed parameter in the model and that the index estimates the decrease in chi-square that would occur if the parameter is set free to be estimated. This means that there is a modification index associated with every path missing in the original path diagram and that it measures the improvement in fit to be expected if a corresponding path is added to the model. Only large modification indices are of interest, so LISREL 8 displays only modification indices larger than 7.882, which is the 99.5 percentile of the chi-square distribution with one degree of freedom. See Section 3.12.2 on how to change this value. To see *all non-zero* modification indices, press **A** (for **A**ll Modification Indices).

This key works as a toggle between *all* and *large* modification indices.

The largest modification index in the figure is 24.7 for the path from Visual to S-C CAPS. If it is difficult to see the value of the modification index or to determine which path it belongs to, several options are available:

- One can zoom and enlarge a smaller window of the path diagram (see Section 3.9).
- One can temporarily delete some paths in the **M**-diagram until the one associated with 24.7 is uniquely identified. In this case, it is sufficient to delete the path from Verbal to S-C CAPS to see that 24.7 is really associated with the path from Visual to S-C CAPS. (See Section 3.5 on how to add or delete a path.)
- While looking at a path diagram, one can display the output file and then go back to the path diagram.

To illustrate the third alternative in this example, press **D** to display the output file and find the section of the output file which looks like the following.

```
        THE MODIFICATION INDICES SUGGEST TO ADD THE
     PATH TO    FROM     DECREASE IN CHI-SQUARE    NEW ESTIMATE
     ADDITION   Visual          10.5                   -.37
     COUNTDOT   Verbal          10.1                   -.28
     S-C CAPS   Visual          24.7                    .57
     S-C CAPS   Verbal          10.0                    .26
```

This output shows that the value 24.7 is associated with the path from Visual to S-C CAPS. Furthermore, this tells us that the parameter associated with this path should be expected to be in the vicinity of .57 if it is included in the model and estimated. In the **M**-diagram, this information is displayed as + in the upper right corner of the modification index. The + indicates that the estimated value of the parameter will be larger if it is set free. A - in the lower right corner of the modification index indicates that the parameter will be smaller if it is set free to be estimated. Press the **Esc** key to get back to the **M**-diagram.

3.5 Adding or Deleting a Path

Adding a path to the model can be done in any **E**-, **T**-, or **M**-diagram. Note that adding a parameter in the model corresponds to adding a path in the

E- or T-diagram or deleting a path in the M-diagram. In the following, we will add a path in the E-diagram.

Continuing the example of the previous section, press **B** (for **B**asic Model) or **X** (for **X**-Measurement Relationships) and then **E** (for Parameter **E**stimates) to get the estimates. You can now use the arrow keys on your keyboard to define the paths in the model as follows. Press any arrow key on the keyboard. The rectangle or ellipse where the cursor is located will appear filled with one color. Use the arrow keys to put the cursor on Visual and press the **Enter** key. Next, put the cursor on S-C CAPS and press the **Enter** key. If you make a mistake, press the **Esc** key before you press the **Enter** key and start over again. The path from Visual to S-C CAPS will now appear in the path diagram with 0.565 attached to it. This is a predicted estimate; the model has not yet been estimated. Before proceeding to re-run the model, it may be instructive to verify that the same path that was added in the E-diagram has been added in the T-diagram, and that this path has disappeared from the M-diagram. Thus, press **T** (for **T**-values), then **M** (for **M**odification Indices), and then **E** (for Parameter **E**stimates). In the T-diagram the path appears with ?.??? attached to it because its t-value has not yet been determined.

Press function key **F3** to estimate the modified model. Soon, a new path diagram will appear on the screen with estimates attached to paths. The parameter estimate for the path from Visual to S-C CAPS is 0.459. Press **F** (for **F**it Measures) and verify that chi-square for the modified model is 28.862 with 23 degrees of freedom. This indicates that the fit of the modified model is acceptable. Note that the chi-square difference between this and the previous model is given in the second line of fit statistics. The reduction in chi-square is $52.626 - 28.862 = 23.764$, roughly the same as the modification index of 24.7 predicted. Press **T** and verify that the t-value of the added path is "significant." Press **M** and verify that no large modification indices remain. This is indicated by an "empty" path diagram.

Deleting a path is done in a similar way. To illustrate this, we continue the example and delete the two-way arrow between Verbal and Speed. Press **E** (for Parameter **E**stimates) to get the estimates. Move the cursor to Verbal and press the **Enter** key. Then move the cursor to Speed and press the **Enter** key. Note that the two-way arrow from Verbal to Speed has disappeared. Press the function key **F3** to run the modified model.

3.6   Adding an Error Covariance

All error terms are assumed to be uncorrelated by default. Occasionally, one may have a reason to specify that two or more error terms should be correlated. This must always be justified and interpreted substantively. This section illustrates how error covariances can be added in the path diagram.

Consider the input file **EX9A.SPL** for Example 9A (Panel Model for Political Efficacy). Add the line

```
Path Diagram
```

and run the problem. After a while, the basic path diagram will appear on the screen. Press **X** (for **X**-Measurement Relationships), **Y** (for **Y**-Measurement Relationships), **S** (for **S**tructural Relationships), and **R** (for E**R**ror Covariances). Verify that the model has been estimated correctly. Press **F** (for **F**it Measures) and verify that the fit of the model is bad: chi-square with 48 degrees of freedom is 115.89. Press **M** (for **M**odification Indices). You should now see the path diagram with modification indices for error covariances. It is given in Figure 3.9.

The two largest modification indices are 19.5 and 45.1 for the errors of VOTING1 and VOTING2 and for COMPLEX1 and COMPLEX2. Both of them have a + attached to them indicating that the covariance of these error terms will be positive if set free to be estimated. Thus, we wish to allow these errors to be correlated. These error covariances are interpreted as the variances of large specific factors in VOTING and COMPLEX (see Aish & Jöreskog, 1990).

Adding an error covariance is the same as deleting a two-way path in the **M**-diagram. To do so, press the arrow keys until the cursor is on VOTING1, and press the **Enter** key. Then put the cursor on VOTING2 and press the **Enter** key. Note that the two-way arrow between VOTING1 and VOTING2 disappears. Follow the same procedure to delete the arrow between COMPLEX1 and COMPLEX2. Press **E** (for Parameter **E**stimates) and verify that two two-way arrows between VOTING1 and VOTING2 and between COMPLEX1 and COMPLEX2 have been added with new predicted values for them. Press the function key **F3** to run the modified model. The estimates of the error covariances will appear in the **R**-diagram with estimates. Press **T** (for **T**-values) to verify that both error covariances are significant. Press **M** (for **M**odification indices), then **X** (for

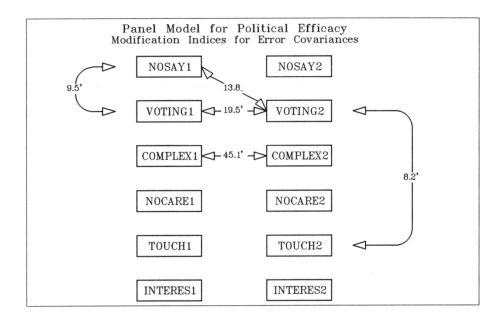

Figure 3.9   R-diagram with Modification Indices for EX9A.SPL

X-Measurement Relationships), **Y** (for **Y**-Measurement Relationships), **S** (for **S**tructural Relationships), and **R** (for E**R**ror Covariances) and verify that no large modification indices remain. This is indicated by an "empty" path diagram.

3.7   Relaxing an Equality Constraint

Example 7 of Chapter 1 and several examples of Chapter 2 involve equality constraints in the sense that two or more parameters in the model are set equal either within a group as in Example 7 or across groups as in Example 10. How to relax such equality constraints at run time will now be described.

Consider Example 10 and input file **EX10C.SPL**. This example has two groups and a model with two indicators of each of two latent variables. The loadings and the error variances are free in each group, but the correlation between the two latent variables are constrained to be equal across groups. The model has a chi-square of 4.03 with three degrees of freedom, so the fit of the model is acceptable. Nevertheless, we use this example to illustrate how to relax an equality constraint across groups.

Add the line

`Path Diagram`

before the line

`End of Problem`

in the input file **EX10C.SPL**, and run the problem. The basic path diagram with estimated parameters for the first group appears on the screen (see Figure 3.10).

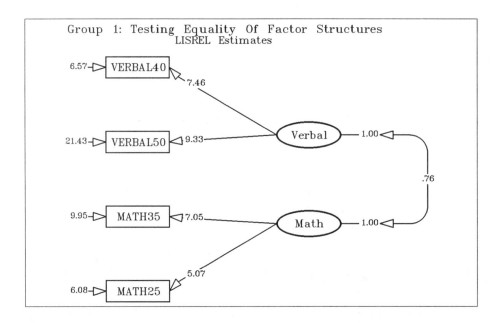

Figure 3.10   B-diagram with Estimates of Group 1 for EX10C.SPL

Press **N** (for Next Group) and the solution for the second group appears on the screen (see Figure 3.11).

Note that the loadings and the error variances are different but the correlation between the two latent variables is the same across groups. Pressing **N** once more brings back the path diagram for the first group.

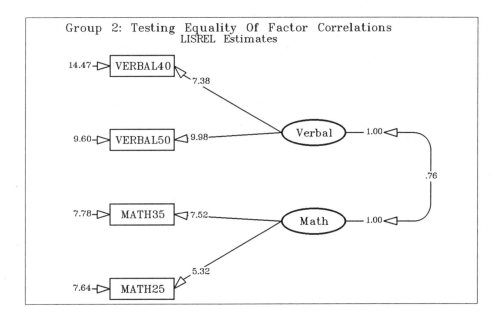

Figure 3.11   B-diagram with Estimates of Group 2 for EX10C.SPL

To relax the equality constraint, try to "delete" the two-way arrow between Verbal and Math in Group 2. That is, with the path diagram for the estimates for the Group 2 on the screen, move the cursor to Verbal and press the **Enter** key, then move the cursor to Math and press the **Enter** key. Note that the two-way arrow does not disappear as it would if the parameter were unconstrained (see Section 3.5). Instead it appears with a .75. This is a predicted value of the correlation between Verbal and Math in Group 2 when it is free from having to be equal to the corresponding value of Group 1. By reestimating the model, a separate factor correlation will be estimated in the two groups. Thus, press the function key **F3** and **N** (for **N**ext Group) twice and verify that the factor correlation is different in the two groups: .78 in Group 1 and .75 in Group 2.

Press **F** (for **F**it Measures) and verify that chi-square is smaller for this model compared to chi-square for Model 10C. In fact, the second line of fit measures says that the chi-square difference is 1.85, indicating that the two correlations are equal.

3.8   Starting with an Empty Path Diagram

It is possible to omit the specification of the model in the input file and specify the model interactively by adding paths on the screen. To do so, just omit the relationships or paths in the input file. We illustrate this by means of three examples.

Consider Example 3 of Chapter 1 and the following input file (**EX3B.SPL**):

```
Union Sentiment of Textile Workers
Variables: y1 = deference (submissiveness) to managers
           y2 = support for labor activism
           y3 = sentiment towards unions
           x1 = years in textile mill
           x2 = age
Observed Variables: y1 - y3 x1 x2

Covariance matrix:
   14.610
   -5.250  11.017
   -8.057  11.087  31.971
   -0.482   0.677   1.559   1.021
  -18.857  17.861  28.250   7.139 215.662

Sample Size 173

Y-Variables: y1 - y3

Path Diagram
End of Problem
```

In this input file we have not specified the model as we did in file **EX3A.SPL**. We have only specified which of the observed variables are y-variables. Recall that the y-variables are the dependent variables. When this problem is run, the program produces an "empty" path diagram on the screen with X-boxes on the left and Y-boxes on the right. It looks like Figure 3.12.

One can now use the arrow keys on the keyboard to define the paths in the model (see Figure 1.3). Press any arrow key on the keyboard. The square at the cursor will appear filled with one color. Use the arrow keys to put the cursor on X1 and press the **Enter** key. Next, put the cursor on Y3 and press the **Enter** key. The path from X1 to Y3 will then be drawn. Proceed in this way to define the other five paths. The general rules are:

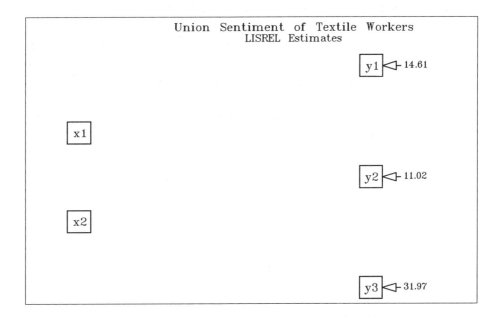

Figure 3.12   Empty B-diagram for EX3B.SPL

□ Put the cursor on the "FROM-VARIABLE" and press the **Enter** key

□ Put the cursor on the "TO-VARIABLE" and press the **Enter** key

Because the program puts the three *y*-variables below each other in the path diagram, it cannot draw the path from Y1 to Y3 as a straight line; instead the path is drawn using several line segments. When all six paths have been drawn, the path diagram looks like Figure 3.13.

To run the program and estimate the model, press the function key **F3**. A path diagram is produced again, this time with estimated coefficients attached to each path. Verify that the results are the same as for Example 3A in Chapter 1.

Next, consider Example 4 of Chapter 1 and the following input file (**EX4B.SPL**):

```
Ability and Aspiration
Observed Variables
'S-C ABIL' PPAREVAL PTEAEVAL PFRIEVAL 'EDUC ASP' 'COL PLAN'
Correlation Matrix From File: EX4.COR
```

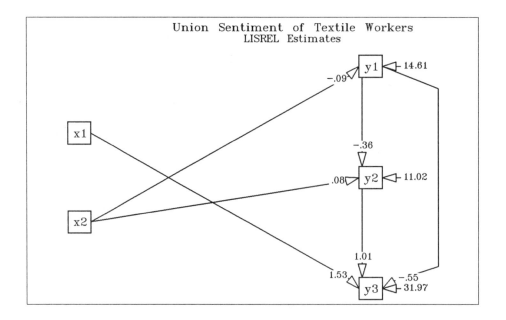

Figure 3.13   B-diagram for EX3B.SPL with Paths Drawn

```
Sample Size: 556
Latent Variables: Ability Aspiratn

Path diagram
End of Problem
```

Unlike file **EX4A.SPL**, this input file specifies no model. Only the observed and latent variables are given. LISREL 8 assumes by default that all latent variables are independent latent variables (Ksi- or ξ-variables) and that all observed variables are x-variables.

Run this file and note what happens on the screen. An "empty" path diagram appears (see Figure 3.14).

The model can now be specified by adding each one-way arrow to the model (see Figure 1.4). As latent independent variables are automatically assumed to be correlated, there is no need to include the two-way arrow between Ability and Aspiration (see Section 6.13, Chapter 6). Press the function key **F3** to run the model and verify the results in Chapter 1.

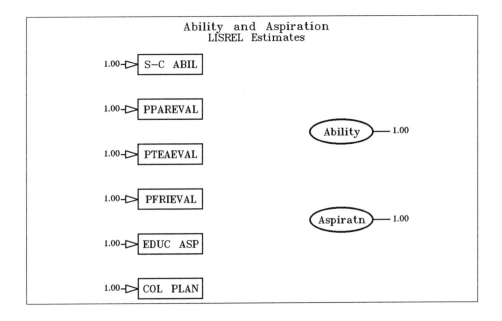

Figure 3.14   Empty B-diagram for EX4B.SPL

Next, consider Example 6 of Chapter 1 and the following input file
(**EX6C.SPL**):

```
Stability of Alienation
Observed Variables
   ANOMIA67   POWERL67   ANOMIA71   POWERL71   EDUC  SEI
Covariance Matrix
   11.834
    6.947    9.364
    6.819    5.091   12.532
    4.783    5.028    7.495    9.986
   -3.839   -3.889   -3.841   -3.625    9.610
   -2.190   -1.883   -2.175   -1.878    3.552    4.503
Sample Size 932
Latent Variables  Alien67 Alien71 Ses
Eta-Variables Alien67 Alien71
Y-Variables ANOMIA67   POWERL67   ANOMIA71   POWERL71
Path Diagram
End of Problem
```

Note the two lines

```
Eta-Variables Alien67 Alien71
Y-Variables ANOMIA67   POWERL67   ANOMIA71   POWERL71
```

The first line specifies which of the latent variables are dependent latent variables and the second specifies which of the observed variables are y-variables, i.e., which of the observed variables specified depend on the dependent latent variables. The difference between **EX6B.SPL** and **EX6C.SPL** is as follows. In **EX6B.SPL**, LISREL 8 can deduce from the relationships in the input file which of the latent variables are dependent and which are independent, as well as which of the observed variables are y-variables and which are x-variables. In **EX6C.SPL**, on the other hand, no relationships are specified, so one must instead specify the dependent latent variables (Eta-variables) and the y-variables. The remaining latent variables and observed variables will be regarded as Ksi-variables and x-variables, respectively. If no Eta-variables are given, LISREL 8 assumes that all latent variables are Ksi-variables (as in **EX4B.SPL**). If no y-variables are given, LISREL 8 assumes that all observed variables are x-variables (as in **EX4B.SPL**).

Now, run this problem. An "empty" path diagram will appear on the screen. This one has four columns of variables, with x-variables, Ksi-variables, Eta-variables, and y-variables, from left to right. Next, specify the model as in Figure 1.6 by adding one path at a time. Altogether there are nine one-way paths to be added. To add the two two-way paths, you must first press **R** (for ERror Covariances) to get the path diagram for correlated error terms. Then the two-way paths can be added as before. Press **B** to get back to the basic path diagram. When all paths have been added, the screen looks like Figure 3.15.

Press the function key **F3** to estimate the model. Verify that the results are the same as for **EX6A.SPL**.

3.9 Zooming

When a path diagram is visible on the screen, any rectangular part of it may be selected for magnification to achieve better visibility. This is particularly useful if the numbers attached to the arrows are not clearly visible. To do so, press **Z** (for **Z**oom) and a rectangular area (window) will appear on the screen. The cursor-control or arrow keys may be used to

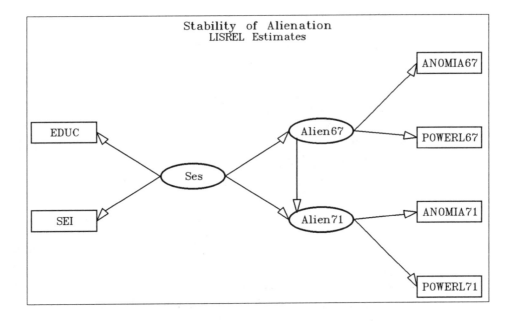

Figure 3.15   B-diagram for EX6C.SPL with Paths Drawn

move the whole window up, down, left or right. To magnify the window, use the Ctrl–5 key combination (hold down the Ctrl key and press the 5 key on the numeric keypad) as a toggle for positive or negative magnification, and use Ctrl and arrow key combinations to move each edge of the window independently. Press the **Enter** key to see the magnified window on the screen. Also, for the magnified window, the **E**, **T**, and **M** keys may be used to switch between estimates, *t*-values, and modification indices. To get back to the original path diagram, press **Z** again.

3.10   Printing a Path Diagram

Any of the path diagrams, *excluding enlarged windows*, may be printed by pressing **P** *when the path diagram is visible on the screen*. This produces a path diagram ten centimeters wide and eight centimeters high, centered on the page both horizontally and vertically. One can control the size of the picture and the location of the upper left corner by including the parameters

```
... SIze=sx,sy [in]   UL=x,y
```

on the Path Diagram line, where

- ❑ x is the distance of the upper left corner of the picture from the left edge of the paper
- ❑ y is the distance of the upper left corner of the picture from the top edge of the paper
- ❑ sx is the horizontal size of the picture
- ❑ sy is the vertical size of the picture

Distances are in centimeters by default (decimals are permitted). To use distances in inches, include IN on the command line. The default values for UL are 0,0 (this is the uppermost leftmost point on the paper which the printer can print, not necessarily the upper left corner of the paper). The default values for SI are 10,8, i.e., the default size is 10 by 8 centimeters.

For example,

```
Path Diagram UL=1.5,2.5 SI=6,8 IN
```

produces a picture of size six by eight inches with the upper left corner 1.5 inches right and 2.5 inches down on the paper.

3.11   Saving the Path Diagram

The path diagram for any model can be saved and reproduced again without actually running LISREL 8 to reestimate the model. This is useful if you want to study the path diagrams on the screen at a later occasion or print them on another printer. It is done as follows.

LISREL 8 automatically saves information about the last model estimated in a file with the same name as the input file but with suffix .PDM (default name INPUT.PDM). The result is an ASCII file containing all information necessary to reproduce all the fifteen different path diagrams for a single model. For multiple groups, all the fifteen path diagrams can be obtained for each group. To produce these path diagrams, one must execute a program called PATHDIAG which does all the drawing. Type the command line

```
PATHDIAG  filename
```

The *filename* is understood to have suffix .PDM unless another suffix is specified. If no filename is given, the file INPUT.PDM will be used.

Watch what happens on the screen. The basic path diagram with parameter estimates appears first. The **S, X, Y, R, B, N, E, T, M** keys work as before, and any path diagram can be printed by pressing **P** as before. *But the model cannot be changed by adding or deleting paths and the model cannot be reestimated.* PATHDIAG is only for viewing and printing.

For printing, the UL and SI keywords defined in the previous section can be entered on the PATHDIAG command line as, for example,

```
PATHDIAG  filename -UL=1.5, 2.5 -SI=6,8 IN
```

If several path diagrams are printed from the same run of PATHDIAG, the same UL and SI keywords apply to all path diagrams. To print path diagrams with different UL and SI keyword values, one must run PATHDIAG for each path diagram. The .PDM file is not changed by PATHDIAG, although it is modified (updated) for each new model estimated with LISREL 8.

3.12 Significance Levels for *t*-Values and Modification Indices

3.12.1 *t*-Values

t-values are given in the output file and in the **T**-diagram for every estimated parameter. In the path diagram, large *t*-values are written in the foreground color (default = white) and small *t*-values in another color (default = red). A threshold value determines which *t*-values will be regarded as large or small. This value is specified as the "significance level" α for the standard normal distribution. *t*-values smaller than $\Phi^{-1}(\alpha/2)$ and larger than $\Phi^{-1}(1 - \alpha/2)$ will be written in the foreground color, where Φ^{-1} is the inverse of the standard normal distribution function. The value of α is specified as a percentage. The default value of α is 5 percent, which means that *t*-values smaller than 1.96 in magnitude will be in the other color. To change the default value of α to one percent, say, put tv=1 on the Path Diagram line.

3.12.2   Modification Indices

Only large modification indices are given in the output file and written
in the **M**-diagram. A threshold value also determines which modification indices will be regarded as large. This value is specified as the α-
"significance level" for the chi-square distribution with one degree of freedom. Modification indices larger than $\Psi^{-1}(1-\alpha)$ will be considered large
and therefore drawn in the path diagram, where Ψ^{-1} is the inverse of the
chi-square distribution function with one degree of freedom. Again, α is
specified as a percentage. The default value for α is 0.5 percent, which
means that modification indices larger than 7.882 are considered large.
To change the default value of α to five percent, say, put `mi=5` on the `Path`
`Diagram` command line.

Example: The following line

```
Path Diagram  tv = 1 mi = 5
```

will set $\alpha = 1\%$ for t-values and $\alpha = 5\%$ for modification indices.

3.13   Summary of Keys

Here is a summary of the different keys and their functions. (on non-US
keyboards press the Alt key together with the key):

B Basic model

X Measurement model for **x**

Y Measurement model for **y**

S Structural equation model

R Variances and covariances of error terms

N Next group

E Estimates

T t-values

M Modification indices

A Toggle between LARGE and ALL modification indices

F Display the fit measures - any key to return to path diagram

D Display the output file - **Esc** to return to path diagram

Z Zoom a window

P Print the path diagram

F1 Help

F3 Reestimate the model

Q Quit looking at path diagrams

4 FITTING AND TESTING

Structural equation models are often used to test a theory about relationships between theoretical constructs. In this chapter we give a short and non-technical account of the main issues involved in the translation of a theory to a structural equation model and the fitting and testing of this model by empirical data. For a fuller account of these issues, see Bollen (1989a), Jöreskog (1993) and other chapters in Bollen & Long (1993). In particular, we discuss issues of model evaluation and assessment of fit that we have only touched upon in the examples of the previous chapters.

4.1 From Theory to Statistical Model

Most theories in the social and behavioral sciences are formulated in terms of hypothetical constructs that cannot be observed or measured directly. Examples of such constructs are prejudice, radicalism, alienation, conservatism, trust, self-esteem, discrimination, motivation, ability, and anomia. The measurement of a hypothetical construct is accomplished indirectly through one or more observable indicators, such as responses to questionnaire items, that are *assumed* to represent the construct adequately .

In theory, the researcher defines the hypothetical constructs by specifying the dimensions of each (see, e.g., Bollen, 1989a, pp. 179–184). The theory further specifies how the various constructs are postulated to be interrelated. This includes first the classification of the constructs as dependent (caused, criterion, endogenous) or independent (causal, explanatory, exogenous). Secondly, for each dependent construct, the theory specifies which of the other constructs it is postulated to depend on. The theory may also include a statement about the sign and/or relative size of the direct effect of one construct on another.

Since the theoretical constructs are not observable, the theory cannot be tested directly. All one can do is examine the *theoretical validity* of the postulated relationships in a given context. Before the theory can be tested empirically, a set of observable indicators must be defined for each dimension of each construct. There must be clear rules of correspondence between the indicators and the constructs, such that each construct and dimension is distinct (Costner, 1969).

The theoretical relationships between the constructs constitute the structural equation part of the model, and the relationships between the observable indicators and the theoretical constructs constitute the measurement part of the model. In order to test the conceptual model, each of these parts must first be formulated as a *statistical* model.

The statistical model requires the specification of the form of the relationship, linear or nonlinear. The structural relationships are usually assumed to be linear but nonlinear models have also been proposed (see e.g., Kenny & Judd, 1984). If the observed indicators are continuous, i.e., measured on interval or ratio scales, the measurement equations are also assumed to be linear, possibly after a logarithmic, exponential, or other nonlinear type of transformation of the observable indicators. If the observed indicators are ordinal, one usually assumes that there is a continuous variable underlying each ordinal variable and formulates the measurement model in terms of these underlying variables (see e.g., Olsson, 1979, Muthén, 1984, and Jöreskog & Sörbom, 1988).

It is not expected that the relationships in the model are exact deterministic relationships. Most often, the independent constructs in the model account for only a fraction of the variation and covariation in the dependent constructs, because there may be many other variables that are associated with the dependent constructs, but are not included in the model. The aggregation of all such omitted variables is represented in the model by a set of stochastic error terms, one for each dependent construct. By definition, these error terms represent the variation and covariation in the dependent constructs left unaccounted for by the independent constructs. *The fundamental assumption in structural equation models is that the error term in each relationship is uncorrelated with all the independent constructs.* Studies should be planned and designed, and variables should be chosen so that this is the case. Failure to do so will lead to biased and inconsistent estimates (omitted variable bias) of the structural coefficients in the linear equations and thus invalidate the testing of the theory. Omitted variables bias is one of the most difficult specification errors to test.

The relationships in the measurement model contain also stochastic error terms that are usually interpreted to be the *sum* of specific factors and random measurement errors in the observable indicators. It is only in specially designed studies such as panel models or multitrait-multimethod designs that one can separate these two components of the error terms. In cross-sectional studies, the error terms should be uncorrelated from one indicator to another. This is part of the definition of indicators of a construct. Because, if the error terms for two or more indicators correlate, it means that the indicators measure *something else* or *something in addition* to the construct they are supposed to measure. If this is the case, the meaning of the construct and its dimensions may be different from what is intended. It is a widespread misuse of structural equation modeling to include correlated error terms[1] in the model for the sole purpose of obtaining a better fit to the data. Every correlation between error terms must be justified and interpreted substantively.

The testing of the structural model, i.e., the testing of the initially specified theory, may be meaningless unless it is first established that the measurement model holds. If the chosen indicators for a construct do not measure that construct, the specified theory must be modified before it can be tested. Therefore, the measurement model should be tested before the structural relationships are tested. It may be useful to do this for each construct separately, then for the constructs taken two at a time, and then for all constructs simultaneously. In doing so, one should let the constructs themselves be freely correlated, i.e., the covariance matrix of the constructs should be unconstrained.

---

[1]We do not include here the situation when error terms correlate for the same variable over time, or other designed conditions due to specific factors or method factors (see Aish & Jöreskog, 1990).

4.2   Nature of Inference

Once a model has been carefully formulated, it can be confronted with empirical data and, if all assumptions hold, various techniques for covariance structure analysis, briefly reviewed in the next section, can be used to test if the model is consistent with the data.

The inference problem is of the following general form. Given assumptions A, B, C, ..., test model M against a more general model M_A. Most researchers will take assumptions A, B, C, ..., for granted and proceed to test M formally. But the assumptions should be checked before the test is made, whenever this is possible.

If the model is rejected by the data, the problem is to determine what is wrong with the model and how the model should be modified to fit the data better.

If the model *fits* the data, it does not mean that it is *the "correct" model or even the "best" model*. In fact, there can be many *equivalent* models, all of which will fit the data equally well as judged by any goodness-of-fit measure. This holds not just for a particular data set but for any data set. The direction of causation and the causal ordering of the constructs cannot be determined by the data[2]. To conclude that the fitted model is the "best," one must be able to exclude all models equivalent to it on logical or substantive grounds. For a good discussion of causal models, see Bollen (1989a, Chapter 3). Further discussion of equivalent models is found in Stelzl (1986), Jöreskog & Sörbom (1989, 1990), and Lee & Hershberger (1990).

Even outside the class of models equivalent to the given model, there may be many models that fit the data almost as well as the given model. Covariance structure techniques can be used to distinguish sharply between models that fit the data *very badly* and those that fit the data *reasonably well*. To discriminate further between models that fit the data almost equally well requires a careful power study (see Satorra & Saris, 1985, Matsueda & Bielby, 1986, and Jöreskog & Sörbom, 1989, pp. 217–221).

---

[2]Different equivalent models will give different parameter estimates, and some may give estimates that are not meaningful. This fact may be used to distinguish some equivalent models from others.

To be strictly testable, the theory should be overidentified in the sense that the structural equation part of the model is overidentified. If the covariance matrix of the construct variables is unconstrained by the model, any test of the model is essentially just a test of the measurement models for the indicators of the constructs. Given that the measurement models hold, the only way to test a saturated structural equation model is to examine whether estimated parameters agree with a priori specified signs and sizes and whether the strengths of the relationships are sufficiently large.

4.3 Fitting and Testing a Covariance Structure

We distinguish between three situations:

SC In a *strictly confirmatory* situation the researcher has formulated one single model and has obtained empirical data to test it. The model should be accepted or rejected.

AM The researcher has specified several *alternative models* or *competing models* and on the basis of an analysis of a single set of empirical data, one of the models should be selected.

MG The researcher has specified a tentative initial model. If the initial model does not fit the given data, the model should be modified and tested again using the same data. Several models may be tested in this process. The goal may be to find a model which not only fits the data well from a statistical point of view, but also has the property that every parameter of the model can be given a substantively meaningful interpretation. The re-specification of each model may be theory-driven or data-driven. Although a model may be tested in each round, the whole approach is *model generating* rather than model testing.

In practice, the **MG** situation is by far the most common. The **SC** situation is very rare because few researchers are content with just rejecting a given model without suggesting an alternative model. The **AM** situation is also rare because researchers seldom specify the alternative models a priori.

We consider the **SC** situation in this section, the **AM** situation in Section 4.4, and the **MG** situation in Section 4.5.

Once the relationships in the theoretical model have been translated into a statistical model for a set of linear stochastic equations among random observable variables (the indicators) and latent variables (the theoretical constructs), the model can be estimated and tested on the basis of empirical data using statistical methods.

A LISREL model may be estimated by seven different methods:

- Instrumental Variables (IV)
- Two-Stage Least Squares (TSLS)
- Unweighted Least Squares (ULS)
- Generalized Least Squares (GLS)
- Maximum Likelihood (ML)
- Generally Weighted Least Squares (WLS)
- Diagonally Weighted Least Squares (DWLS)

For a precise definition of these methods see Jöreskog & Sörbom (1989). Under general assumptions, all seven methods give consistent estimates of parameters. This means that they will be close to the true parameter values in large samples (assuming, of course, that the model is correct). The seven types of estimates differ in several respects. The TSLS and IV methods are procedures that are non-iterative and very fast. They are used to compute starting values for the other methods but can also be requested as final estimates. The ULS, GLS, ML, WLS, and DWLS estimates are obtained by means of an iterative procedure that minimizes a particular fit function by successively improving the parameter estimates. WLS requires an estimate of the asymptotic covariance matrix of the sample variances and covariances or correlations being analyzed. Similarly, DWLS requires an estimate of the asymptotic variances of the sample variances and covariances or correlations being analyzed. These asymptotic variances and covariances are obtained by PRELIS (Jöreskog & Sörbom, 1988), which saves them in a file to be read by LISREL. See Section 6.14.4 of Chapter 6 on how to specify the method of estimation.

The statistical model and its assumptions imply a covariance structure $\Sigma(\theta)$ for the observable random variables, where $\theta = (\theta_1, \theta_2, \ldots, \theta_t)$ is a vector of parameters in the statistical model. The testing problem is now conceived of as testing the model $\Sigma(\theta)$. It is assumed that the empirical data is a random sample of size N of cases (individuals) on which the observable variables have been observed or measured. From this data a sample covariance matrix $\mathbf{S}$ is computed, and it is this matrix that is used to fit the model to the data and to test the model.

The model is fitted by minimizing a fit function $F[\mathbf{S}, \boldsymbol{\Sigma}(\boldsymbol{\theta})]$ of $\mathbf{S}$ and $\boldsymbol{\Sigma}(\boldsymbol{\theta})$ which is non-negative and zero only if there is a perfect fit, in which case $\mathbf{S}$ equals $\boldsymbol{\Sigma}(\boldsymbol{\theta})$. Although the family of fit functions F includes all the fit functions that are used in practice, e.g., ULS, GLS, ML, DWLS, and WLS, see Jöreskog & Sörbom (1989), we will be concerned only with ML, GLS and WLS here.

Suppose that $\mathbf{S}$ converges in probability to $\boldsymbol{\Sigma}_0$ as the sample size increases, and let $\boldsymbol{\theta}_0$ be the value of $\boldsymbol{\theta}$ that minimizes $F[\boldsymbol{\Sigma}_0, \boldsymbol{\Sigma}(\boldsymbol{\theta})]$. We say that the model holds if $\boldsymbol{\Sigma}_0 = \boldsymbol{\Sigma}(\boldsymbol{\theta}_0)$. Furthermore, let $\hat{\boldsymbol{\theta}}$ be the value of $\boldsymbol{\theta}$ that minimizes $F[\mathbf{S}, \boldsymbol{\Sigma}(\boldsymbol{\theta})]$ for the given sample covariance matrix $\mathbf{S}$, and let $n = N - 1$.

To test the model, one may use

$$c = nF[\mathbf{S}, \boldsymbol{\Sigma}(\hat{\boldsymbol{\theta}})] . \tag{4.1}$$

If the model holds and is identified, c is approximately distributed in large samples as χ^2 with $d = s - t$ degrees of freedom, where $s = k(k+1)/2$ and t is the number of independent parameters estimated. To use this test formally, one chooses a significance level α, draws a random sample of observations on $z_1, z_2, \ldots, z_k$, computes $\mathbf{S}$, estimates the model, and rejects the model if c exceeds the $(1 - \alpha)$ percentile of the χ^2 distribution with d degrees of freedom. As will be argued in the next section, this is a highly limited view of the testing problem.

If the model does not hold, $\boldsymbol{\Sigma}_0 \neq \boldsymbol{\Sigma}(\boldsymbol{\theta}_0)$ and c in (4.1) is distributed as non-central χ^2 with $s - t$ degrees of freedom and non-centrality parameter (Browne, 1984)

$$\lambda = nF[\boldsymbol{\Sigma}_0, \boldsymbol{\Sigma}(\boldsymbol{\theta}_0)] , \tag{4.2}$$

an unknown population quantity that may be estimated as

$$\hat{\lambda} = \mathsf{Max}\{(c - d), 0\} , \tag{4.3}$$

cf., Browne & Cudeck (1993). One can also set up a confidence interval for λ. Let $\hat{\lambda}_L$ and $\hat{\lambda}_U$ be the solutions of

$$G(c|\lambda_L, d) = 0.95 \tag{4.4}$$

$$G(c|\lambda_U, d) = 0.05 \,, \tag{4.5}$$

where $G(x|\lambda, d)$ is the distribution function of the non-central chi-square distribution with non-centrality parameter λ and d degrees of freedom. Then $(\hat{\lambda}_L; \hat{\lambda}_U)$ is a 90 percent confidence interval for λ (see Browne & Cudeck, 1993).

Once the validity of the model has been established, one can test structural hypotheses about the parameters $\boldsymbol{\theta}$ in the model such that

- certain θ's have particular values (fixed parameters)
- certain θ's are equal (equality constraints)
- certain θ's are specified linear or nonlinear functions of of other parameters.[3]

Each of these types of hypotheses leads to a model with fewer parameters, u, say, for $u < t$ and with a parameter vector $\boldsymbol{\nu}$ of order $(u \times 1)$ containing a subset of the parameters in $\boldsymbol{\theta}$. In conventional statistical terminology, the model with parameters $\boldsymbol{\nu}$ is called the *null hypothesis* H_0 and the model with parameters $\boldsymbol{\theta}$ is called the *alternative hypothesis* H_1. Let c_0 and c_1 be the value of c for models H_0 or H_1, respectively. The *likelihood ratio test statistic* for testing H_0 against H_1 is then

$$D^2 = c_0 - c_1 \tag{4.6}$$

which is used as χ^2 with $t - u$ degrees of freedom. The degrees of freedom can also be computed as the difference between the degrees of freedom associated with c_0 and c_1.

To use the tests formally, one chooses a significance level α (probability of a type 1 error) and rejects H_0 if D^2 exceeds the $(1 - \alpha)$ percentile of the χ^2 distribution with $t - u$ degrees of freedom.

4.4 Selection of One of Several a priori Specified Models

The previous sections have focused on the testing of a single specified model including the checking of the assumptions on which the test is

---

[3]Linear and nonlinear constraints cannot be tested using the SIMPLIS language; the LISREL language is required for these tests.

based. In this section, we consider the **AM** situation, in which the researcher has specified a priori a number of alternative or competing models $M_1, M_2, \ldots, M_k$, say, and wants to select one of these. Here it is assumed that all models are identified and can be tested with a valid chi-square test as in Section 4.3. It is also assumed that each model gives reasonable results apart from fit considerations. Any model that gives unreasonable results can be eliminated from further consideration.

We begin with the case in which the models are nested so they can be ordered in decreasing number of parameters (increasing degrees of freedom). M_1 is the most flexible model, i.e., the one with most parameters (fewest degrees of freedom). M_k is the most restrictive model, i.e., the one with fewest parameters (largest degrees of freedom). If the models are nested, each model is a special case of the preceding model, i.e., M_i is obtained from M_{i-1} by placing restrictions on the parameters of M_{i-1}.

We assume that model M_1 has been tested by a valid chi-square test (as discussed in previous sections), and found to fit the data and to be interpretable in a meaningful way. We can then test sequentially M_2 against M_1, M_3 against M_2, M_4 against M_3, and so on. Each test can be a likelihood ratio test, a Lagrangian multiplier test, or a Wald test. If M_i is the first model rejected, one takes M_{i-1} to be the "best" model. Under this procedure, the probability of rejecting M_i when M_j is true is not known. There is no guarantee that the tests are independent, not even asymptotically. In general, what is required is a multiple decision procedure that takes explicitly into account the probability of rejecting each hypothesis when a particular one is true. The mathematical theory for such a procedure is so complicated that its practical value is very limited.

Another approach is to compare the models on the basis of some criteria that takes parsimony (in the sense of number of parameters) into account as well as fit. This approach can be used regardless of whether or not the models can be ordered in a nested sequence. Three strongly related criteria have been proposed: the AIC measure of Akaike (1974, 1987), the CAIC by Bozdogan (1987), and the single sample cross-validation index ECVI by Browne & Cudeck (1989). All of these are simple functions of chi-square and the degrees of freedom:

$$\text{AIC} = c + 2t \tag{4.7}$$

$$\text{CAIC} = c + (1 + \ln N)t \tag{4.8}$$

$$\text{ECVI} = (c/n) + 2(t/n) \tag{4.9}$$

where c is the chi-square measure of overall fit of the model given by (4.1), and n and t are as defined in Section 4.3. Although ECVI is quite similar to AIC, the rationale for ECVI is quite different from that of AIC and CAIC. Whereas AIC and CAIC are derived from statistical information theory, the ECVI is a measure of the discrepancy between the fitted covariance matrix in the analyzed sample and the expected covariance matrix that would be obtained in another sample of the same size.

To apply these measures to the decision problem, one estimates each model, ranks them according to one of these criteria and chooses the model with the smallest value. Browne & Cudeck (1993) point out that one can also take the precision of the estimated value of ECVI into account. For example, a 90 percent confidence interval for ECVI is

$$\left(\frac{\hat{\lambda}_L + s + t}{n} \; ; \frac{\hat{\lambda}_U + s + t}{n} \right) . \tag{4.10}$$

For a given data set, n is the same for all models. Therefore, AIC and ECVI will give the same rank order of the models, whereas the rank ordering of AIC and CAIC can differ. To see how much they can differ, consider the case of two nested models M_1 and M_2 differing by one degree of freedom. Let $D^2 = c_2 - c_1$. The conventional testing procedure will reject M_2 in favor of M_1 if D^2 exceeds the $(1 - \alpha)$ percentage point of the chi-square distribution with one degree of freedom. If $\alpha = 0.05$ this will lead to accepting M_1 if D^2 exceeds 3.84. AIC will select M_1 if D^2 exceeds 2, whereas CAIC will select M_1 if D^2 exceeds $1 + \ln N$.

4.5   Model Assessment and Modification

We now consider the **MG** case. The problem is not just to accept or reject a specified model or to select one out of a set of specified models. Rather, the researcher has specified an initial model that is not assumed to hold exactly in the population and may only be tentative. Its fit to the data is to be evaluated and assessed in relation to what is known about the substantive area, the quality of the data, and the extent to which various assumptions are satisfied. The evaluation of the model and the assessment of fit is not entirely a statistical matter. If the model is judged not to be good on substantive or statistical grounds, it should be modified within a class of models suitable for the substantive problem. The goal is to find a model within this class of models that not only fits the data well statistically, taking all aspects of error into account, but that also has the property of every parameter having a substantively meaningful interpretation.

The output from a structural equations program provides much information useful for model evaluation and assessment of fit. It is helpful to classify this information into the three groups

☐ Examination of the solution

☐ Measures of overall fit

☐ Detailed assessment of fit

and to proceed as follows.

1. Examine the parameter estimates to see if there are any unreasonable values or other anomalies. Parameter estimates should have the right sign and size according to theory or a priori specifications. Examine the squared multiple correlation R^2 for each relationship in the model. The R^2 is a measure of the strength of linear relationship. A small R^2 indicates a weak relationship and suggests that the model is not effective.

2. Examine the measures of overall fit of the model, particularly the chi-square (see Section 4.5.1). A number of other measures of overall fit (see Section 4.5.2), all of which are functions of chi-square, may also be used. If any of these quantities indicate a poor fit of the data, proceed with the detailed assessment of fit in the next step.

3. The tools for examining the fit in detail are the residuals and standardized residuals, the modification indices, as well as the expected change (see Section 4.5.3). Each of these quantities may be used to locate the source of misspecification and to suggest how the model should be modified to fit the data better.

4.5.1 Chi-square

In Section 4.3, we regarded c in (4.1) as a chi-square for testing the model against the alternative that the covariance matrix of the observed variables is unconstrained. This is valid if all assumptions are satisfied, if the model holds and the sample size is sufficiently large. In practice it is more useful to regard chi-square as a *measure* of fit rather than as a *test statistic*. In this view, chi-square is a measure of overall fit of the model to the data. It measures the distance (difference, discrepancy, deviance) between the sample covariance (correlation) matrix and the fitted covariance (correlation) matrix. Chi-square is a badness-of-fit measure in the sense that a small chi-square corresponds to good fit and a large chi-square to bad fit. Zero chi-square corresponds to perfect fit. Chi-square is calculated as $N - 1$ times the minimum value of the fit function, where N is the sample size.

Even if all the assumptions of the chi-square test hold, it may not be realistic to assume that the model holds exactly in the population. In this case, chi-square should be compared with a non-central rather than a central chi-square distribution (see Browne, 1984).

4.5.2 Other Goodness-of-Fit Measures

A number of other goodness-of-fit measures have been proposed and studied in the literature. For a summary of these and the rationale behind them, see Bollen (1989a). All of these measures are functions of chi-square.

Goodness-of-Fit Indices

Since chi-square is $N - 1$ times the minimum value of the fit function, chi-square tends to be large in large samples if the model does not hold. A number of goodness-of-fit measures have been proposed to eliminate or reduce its dependence on sample size. This is a hopeless task, however, because even though a measure does not depend on sample size explicitly in its calculation, its sampling distribution will depend on N. The goodness-of-fit measures GFI and AGFI of Jöreskog & Sörbom (1989) (cf., Tanaka & Huba, 1985) do not depend on sample size explicitly and measure how much *better* the model fits as compared to no model at all. For a specific class of models, Maiti & Mukherjee (1990) demonstrate that there is an exact monotonic relationship between GFI and chi-square.

The goodness-of-fit index (GFI) is defined as

$$\text{GFI} = 1 - \frac{F[\mathbf{S}, \mathbf{\Sigma}(\hat{\theta})]}{F[\mathbf{S}, \mathbf{\Sigma}(\mathbf{0})]} \ . \tag{4.11}$$

The numerator in (4.11) is the minimum of the fit function after the model has been fitted; the denominator is the fit function before any model has been fitted, or when all parameters are zero.

The goodness-of-fit index adjusted for degrees of freedom, or the adjusted GFI, AGFI, is defined as

$$\text{AGFI} = 1 - \frac{k(k+1)}{2d}(1 - \text{GFI}) \ , \tag{4.12}$$

where d is the degrees of freedom of the model. This corresponds to using mean squares instead of total sums of squares in the numerator and denominator of $1 - \text{GFI}$. Both of these measures should be between zero and one, although it is theoretically possible for them to become negative. This should not happen, of course, since it means that the model fits worse than no model at all.

Fit Measures Based on Population Error of Approximation

The use of chi-square as a central χ^2-statistic is based on the assumption that the model holds exactly in the population. As already pointed out, this may be an unreasonable assumption in most empirical research. A consequence of this assumption is that models which hold approximately in the population will be rejected in large samples. Browne & Cudeck (1993) proposed a number of fit measures which take particular account of the error of approximation in the population and the precision of the fit measure itself. They define an estimate of the *population discrepancy function* (PDF) as (cf., McDonald, 1989)

$$\hat{F}_0 = \text{Max}\{\hat{F} - (d/n), 0\} \ , \tag{4.13}$$

where $\hat{F}$ is the minimum value of the fit function, $n = N - 1$ and d is the degrees of freedom, and suggest the use of a 90 percent confidence interval

$$(\hat{\lambda}_L/n; \hat{\lambda}_U/n) \tag{4.14}$$

to assess the error of approximation in the population.

Since $\hat{F}_0$ generally decreases when parameters are added in the model, Browne & Cudeck (1993) suggest using Steiger's (1990) Root Mean Square Error of Approximation (RMSEA)

$$\epsilon = \sqrt{\hat{F}_0/d} \qquad (4.15)$$

as a measure of *discrepancy per degree of freedom*. Browne & Cudeck (1993) suggest that a value of 0.05 of ϵ indicates a close fit and that values up to 0.08 represent reasonable errors of approximation in the population. A 90 percent confidence interval of ϵ and a test of $\epsilon < 0.05$ give quite useful information for assessing the degree of approximation in the population. A 90 percent confidence interval for ϵ is

$$\left(\sqrt{\frac{\hat{\lambda}_L}{nd}} ; \sqrt{\frac{\hat{\lambda}_U}{nd}} \right) \qquad (4.16)$$

and the P-value for test of $\epsilon < 0.05$ is calculated as

$$P = 1 - G(c|0.0025nd, d) \qquad (4.17)$$

(see Browne & Cudeck, 1993).

Information Measures of Fit

One disadvantage with chi-square in comparative model fitting is that it always decreases when parameters are added to the model. Therefore there is a tendency to add parameters to the model so as to make chi-square small, thereby capitalizing on chance and ending with a model containing nonsense parameters. A number of measures of fit have been proposed that take parsimony (in the sense of as few parameters as possible) as well as fit into account. These approaches try to deal with this problem by constructing a measure which ideally first decreases as parameters are added and then has a turning point such that it takes its smallest value for the "best" model and then increases when further parameters are added. The AIC and CAIC measures as well as the ECVI discussed previously belong to this category.

These measures for the estimated model may be compared with the same measures for the "independence" model, i.e., the model in which all observed variables are uncorrelated (if the variables are normally distributed they are independent and their covariance matrix is diagonal; this model has k parameters and $k(k-1)/2$ degrees of freedom, where k is the number of observed variables). It will also be useful to compare these measures for the model estimated with those of the saturated model with $k(k+1)/2$ parameters and zero degrees of freedom. The chi-square for the independence model should also be considered. This provides a test of the hypothesis that the observed variables are uncorrelated. If this hypothesis is not rejected, it is pointless to model the data.

Other Fit Indices

Another class of fit indices measures how much *better* the model fits as compared to a baseline model, usually the independence model. The first indices of this kind were developed by Tucker & Lewis (1973) and Bentler & Bonett (1980) (NNFI, NFI). Other variations of these have been proposed and discussed by Bollen (1986, 1989a,b) (RFI, IFI) and Bentler (1990) (CFI). These indices are supposed to lie between 0 and 1, but values outside this interval can occur, and, since the independence model almost always has a huge chi-square, one often obtains values very close to 1. James, Mulaik, & Brett (1982, p.155) suggest taking parsimony (degrees of freedom) into account and define a parsimony normed fit index (PNFI), and Mulaik, *et al.* (1989) suggest a parsimony goodness-of-fit index (PGFI)

For completeness, we give the definition of each of these measures here. Let F be the minimum value of the fit function for the estimated model, let F_i be the minimum value of the fit function for the independence model, and let d and d_i be the corresponding degrees of freedom. Furthermore, let $f = nF/d$, $f_i = nF_i/d_i$, $\tau = \max(nF - d, 0)$, and $\tau_i = \max(nF_i - d_i, nF - d, 0)$. Then

$$\text{NFI} = 1 - F/F_i \qquad (4.18)$$

$$\text{PNFI} = (d/d_i)(1 - F/F_i) \qquad (4.19)$$

$$\text{NNFI} = \frac{f_i - f}{f_i - 1} \qquad (4.20)$$

$$\text{CFI} = 1 - \tau/\tau_i \qquad (4.21)$$

$$\text{IFI} = \frac{nF_i - nF}{nF_i - d} \qquad (4.22)$$

$$\text{RFI} = \frac{f_i - f}{f_i} \qquad (4.23)$$

$$\text{PGFI} = (2d/k(k+1))\text{GFI} \qquad (4.24)$$

Hoelter (1983) proposed a critical N (CN) statistic:

$$\text{CN} = \frac{\chi^2_{1-\alpha}}{F} + 1 , \qquad (4.25)$$

where $\chi^2_{1-\alpha}$ is the $1 - \alpha$ percentile of the chi-square distribution. This is the sample size that would make the obtained chi-square just significant at the significance level α. For a discussion of this statistic, see Bollen & Liang (1988) and Bollen (1989a).

4.5.3 Detailed Assessment of Fit

If, on the basis of overall measures of fit or other considerations, it is concluded that the model does not fit sufficiently well, one can examine the fit more closely to determine possible sources of the lack of fit. For this purpose, fitted and standardized residuals and modification indices are useful.

Fitted and Standardized Residuals

A residual is an observed minus a fitted covariance (variance). A standardized residual is a residual divided by its estimated standard error. There are such residuals for every pair of observed variables. Fitted residuals depend on the unit of measurement of the observed variables. If the variances of the variables vary considerably from one variable to another, it is rather difficult to know whether a fitted residual should be considered large or small. Standardized residuals, on the other hand, are independent of the units of measurement of the variables. In particular, standardized residuals provide a "statistical" metric for judging the size of a residual.

A large positive residual indicates that the model underestimates the covariance between the two variables. On the other hand, a large negative residual indicates that the model overestimates the covariance between the variables. In the first case, one should modify the model by adding paths which could account for the covariance between the two variables better. In the second case, one should modify the model by eliminating paths that are associated with the particular covariance.

All the standardized residuals may be examined collectively in two plots: a stemleaf plot and a Q-plot. A good model is characterized by a stemleaf plot in which the residuals are symmetrical around zero, with most in the middle and fewer in the tails. An excess of residuals on the positive or negative side indicates that residuals may be systematically under- or overestimated in the above sense. In the Q-plot, a good model is characterized by points falling approximately on a 45° line. Deviations from this pattern are indicative of specification errors in the model, nonnormality in the variables or of nonlinear relationships among the variables. In particular, standardized residuals that appear as outliers in the Q-plot are indicative of a specification error in the model.

Modification Index

A modification index (Sörbom 1989) may be computed for each fixed and constrained parameter in the model. Each such modification index measures how much chi-square is expected to decrease if this particular parameter is set free and the model is reestimated. Thus, the modification index is approximately equal to the difference in chi-square between two models in which one parameter is fixed or constrained in one model and free in the other, all other parameters being estimated in both models. The largest modification index shows the parameter that improves the fit most when set free.

Associated with each modification index, there is an expected parameter change (EPC) (Saris, Satorra, & Sörbom, 1987; Jöreskog & Sörbom, 1989). This measures how much the parameter is expected to change, in the positive or negative direction, if it is set free. If the units of measurement in observed and/or latent variables are of no particular interest, Kaplan (1989) suggested using a scalefree variant SEPC of EPC.

Modification indices are used in the process of model evaluation and modification in the following way. If chi-square is large relative to the degrees of freedom, one examines the modification indices and relaxes the parameter with the largest modification index *if this parameter can be interpreted substantively*. If it does not make sense to relax the parameter with the largest modification index, one considers the second largest modification index, etc. If the signs of certain parameters are specified a priori, positive or negative, the expected parameter changes associated with the modification indices for these parameters can be used to exclude models with parameters having the wrong sign.

4.5.4 Strategy of Analysis

A suitable strategy for data analysis in the **MG** case may be the following.

1. Specify an initial model on the basis of substantive theory, stated hypotheses, or at least some tentative ideas of what a suitable model should be.

2. Estimate the measurement model for each construct separately, then for each pair of constructs, combining them two by two. Then estimate the measurement model for all the constructs without constraining the covariance matrix of the constructs. Finally, estimate the structural equation model for the constructs jointly with the measurement models.

3. For each model estimated in Step 2, evaluate the fit as described in the previous subsections. In particular, pay attention to chi-square, standard errors, t-values, standardized residuals, and modification indices. If chi-square is large relative to the degrees of freedom, the model must be modified to fit the data better. For model modification, follow the hints in Section 4.5.3. If chi-square is small relative to the degrees of freedom, the model is overfitted and parameters with very large standard errors (very small t-values) could possibly be eliminated. If chi-square is in the vicinity of the degrees of freedom, the model may be acceptable, but examine the estimated solution to see if there are any unreasonable values or other anomalies.

   For each model estimated, if this step leads to a modified model, repeat this step on each modified model.

4. Hopefully, the last model estimated in the previous step is one which fits the data of the sample reasonably well and in which all parameters are meaningful and substantively interpretable. However, this does not necessarily mean that it is the "best" model, because its results may have been obtained to some extent by "capitalizing on chance." The model modification process may have generated several "reasonable models" which should be cross-validated on independent data. If no independent sample is available but the initial

sample is large, one may consider splitting the sample into two sub-samples and use one (the calibration sample) for exploration as described previously and the other (the validation sample) for cross-validation. The cross-validation is done by computing a validation index (Cudeck & Browne, 1983) for each model. The validation index is a measure of the distance (difference, discrepancy, deviance) between the fitted covariance matrix in the calibration sample and the sample covariance matrix of the validation sample. The model with the smallest validation index is the one which is expected to be most stable in repeated samples. If the smallest validation index occurs for the best fitted model in the calibration sample it is good. If the smallest validation index occurs for some of the other models that have been fitted, one must make a decision on substantive grounds, which of the two models to retain.

4.6 Illustration

To illustrate all the goodness-of-fit statistics, we return to Example 5A of Chapter 1. The fit statistics are given in the output as

```
       CHI-SQUARE WITH 24 DEGREES OF FREEDOM = 52.626 (P = 0.000648)
             ESTIMATED NON-CENTRALITY PARAMETER (NCP) = 28.626
         90 PERCENT CONFIDENCE INTERVAL FOR NCP = (11.415 ; 53.568)


                   MINIMUM FIT FUNCTION VALUE = 0.365
            POPULATION DISCREPANCY FUNCTION VALUE (F0) = 0.199
          90 PERCENT CONFIDENCE INTERVAL FOR F0 = (0.0793 ; 0.372)
         ROOT MEAN SQUARE ERROR OF APPROXIMATION (RMSEA) = 0.0910
       90 PERCENT CONFIDENCE INTERVAL FOR RMSEA = (0.0575 ; 0.124)
           P-VALUE FOR TEST OF CLOSE FIT (RMSEA < 0.05) = 0.0250


             EXPECTED CROSS-VALIDATION INDEX (ECVI) = 0.657
        90 PERCENT CONFIDENCE INTERVAL FOR ECVI = (0.538 ; 0.830)
                   ECVI FOR SATURATED MODEL = 0.625

 CHI-SQUARE FOR INDEPENDENCE MODEL WITH 36 DEGREES OF FREEDOM = 496.218
                     INDEPENDENCE AIC = 514.218
                        MODEL AIC = 94.626
                      SATURATED AIC = 90.000
                    INDEPENDENCE CAIC = 550.009
                       MODEL CAIC = 178.137
                     SATURATED CAIC = 268.953
```

```
          ROOT MEAN SQUARE RESIDUAL (RMR) = 0.0755
                 STANDARDIZED RMR = 0.0755
            GOODNESS OF FIT INDEX (GFI) = 0.928
   ADJUSTED GOODNESS OF FIT INDEX (AGFI) = 0.866
  PARSIMONY GOODNESS OF FIT INDEX (PGFI) = 0.495


               NORMED FIT INDEX (NFI) = 0.894
           NON-NORMED FIT INDEX (NNFI) = 0.907
      PARSIMONY NORMED FIT INDEX (PNFI) = 0.596
          COMPARATIVE FIT INDEX (CFI) = 0.938
          INCREMENTAL FIT INDEX (IFI) = 0.939
             RELATIVE FIT INDEX (RFI) = 0.841


                CRITICAL N (CN) = 118.606
```

The chi-square test of exact fit would reject the model, since the P-value is very small. Following the guidelines of Browne & Cudeck (1993), it is seen that the point estimate of RMSEA is 0.091 and the 90 percent confidence interval is from 0.0574 to 0.124. Since the lower bound is above the recommended value of 0.05, it is concluded that the degree of approximation in the population is too large. So the model is rejected.

It was demonstrated in Chapter 1 how one can use standardized residuals and modification indices to determine the main source of the lack of fit. For the modified model, the fit indices are

```
   CHI-SQUARE WITH 23 DEGREES OF FREEDOM = 28.862 (P = 0.185)
       ESTIMATED NON-CENTRALITY PARAMETER (NCP) = 5.862
     90 PERCENT CONFIDENCE INTERVAL FOR NCP = (0.0 ; 23.758)


              MINIMUM FIT FUNCTION VALUE = 0.200
    POPULATION DISCREPANCY FUNCTION VALUE (F0) = 0.0407
     90 PERCENT CONFIDENCE INTERVAL FOR F0 = (0.0 ; 0.165)
   ROOT MEAN SQUARE ERROR OF APPROXIMATION (RMSEA) = 0.0421
   90 PERCENT CONFIDENCE INTERVAL FOR RMSEA = (0.0 ; 0.0847)
       P-VALUE FOR TEST OF CLOSE FIT (RMSEA ¦ 0.05) = 0.574


       EXPECTED CROSS-VALIDATION INDEX (ECVI) = 0.506
    90 PERCENT CONFIDENCE INTERVAL FOR ECVI = (0.465 ; 0.630)
            ECVI FOR SATURATED MODEL = 0.625

  CHI-SQUARE FOR INDEPENDENCE MODEL WITH 36 DEGREES OF FREEDOM = 496.218
                 INDEPENDENCE AIC = 514.218
                    MODEL AIC = 72.862
```

```
            SATURATED AIC = 90.000
       INDEPENDENCE CAIC = 550.009
             MODEL CAIC = 160.350
         SATURATED CAIC = 268.953


   ROOT MEAN SQUARE RESIDUAL (RMR) = 0.0452
             STANDARDIZED RMR = 0.0452
       GOODNESS OF FIT INDEX (GFI) = 0.958
 ADJUSTED GOODNESS OF FIT INDEX (AGFI) = 0.917
 PARSIMONY GOODNESS OF FIT INDEX (PGFI) = 0.489


          NORMED FIT INDEX (NFI) = 0.942
      NON-NORMED FIT INDEX (NNFI) = 0.980
  PARSIMONY NORMED FIT INDEX (PNFI) = 0.602
     COMPARATIVE FIT INDEX (CFI) = 0.987
     INCREMENTAL FIT INDEX (IFI) = 0.988
       RELATIVE FIT INDEX (RFI) = 0.909
```

Here the point estimate of RMSEA is below 0.05 and the upper confidence limit is only slightly above the value 0.08 suggested by Browne & Cudeck (1993). The P-value for test of close fit is 0.574. Another indication that the model fits well is that the ECVI for the model (0.506) is less than the ECVI for the saturated model (0.625). In fact, the confidence interval for ECVI is from 0.465 to 0.630. We conclude that the model fits well and represents a reasonably close approximation in the population.

5 LISREL OUTPUT

The examples of the previous chapters have demonstrated that the estimated model may be obtained in two different forms, namely as path diagrams on the screen or as equations in the output file, or both. There is still another way, namely in LISREL form. Thus, with SIMPLIS input, we can get an output file either in SIMPLIS format or in LISREL format. The main difference between the two formats is that in SIMPLIS format the estimated model is given in the form of equations, whereas in LISREL format the model is given in terms of parameter matrices. A further important reason for considering output in LISREL format is that it is possible to obtain additional information in the output file that is not available in the SIMPLIS output. While the most essential information about the estimated model is already given in the SIMPLIS output, some users of LISREL 8 may be interested in the additional information that one can get from LISREL output. Understanding the LISREL output may also serve as a first step to learn the LISREL input language (Jöreskog & Sörbom, 1989).

To obtain the output in LISREL format, include the line

```
LISREL Output
```

in the input file. The different kinds of optional output one can ask for are specified by two-character keywords on the same line, for example,

```
LISREL Output: RS MI SS SC EF
```

The LISREL output and the various options are described in this chapter.

5.1 Hypothetical Model

To illustrate the LISREL output and the various options available, we will use the hypothetical model shown in Figure 5.1. This model includes most of the elements of LISREL modeling.

133

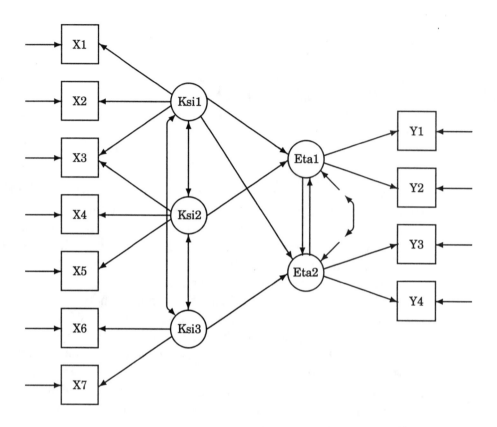

Figure 5.1   Path Diagram for Hypothetical Model

Since this is a hypothetical model we use generic names Y1, Y2, Y3, Y4, X1, X2, ..., X7, to represent the observed variables and Eta1, Eta2, Ksi1, Ksi2, Ksi3, to represent the latent variables. If this were a real model, these names would be replaced by names associated with the content of the variables. It should be emphasized that in choosing the names of the variables, we have already classified each group of variables. However, this is not necessary; any names could have been chosen. The classification of the variables is performed by the program, see Section 5.2. A random sample of 100 observations of the observed variables has been generated artificially and the sample covariance matrix has been saved in the file **EX17.COV**. The SIMPLIS input file for the hypothetical model is (**EX17A.SPL**):

```
Hypothetical Model
Observed Variables: Y1-Y4 X1-X7
Covariance Matrix from File EX17.COV
Sample Size: 100
Latent Variables: Eta1 Eta2 Ksi1-Ksi3

Relationships
    Eta1 = Eta2 Ksi1 Ksi2
    Eta2 = Eta1 Ksi1 Ksi3
Let the Errors of Eta1 and Eta2 Correlate

    Y1 = 1*Eta1
    Y2 = Eta1
    Y3 = 1*Eta2
    Y4 = Eta2

    X1 = 1*Ksi1
    X2 X3 = Ksi1
    X4 = 1*Ksi2
    X3 X5 = Ksi2
    X6 = 1*Ksi3
    X7 = Ksi3

LISREL Output: RS MI SC EF WP
End of Problem
```

Note the following

❑ On the line

```
Observed Variables: Y1-Y4 X1-X7
```

the observed variables Y1, Y2, Y3, Y4, can be defined as Y1-Y4. Similarly, for X1-X7 (see Section 6.3).

❑ In this example we assume that the observed variables are measured in some well-defined units of measurements that we want to retain in the analysis. Consequently, we analyze the covariance matrix of the observed variables rather than the correlation matrix. Furthermore, by using observed variables as reference variables for the latent variables, these will also have units of measurements that are interpretable. The reference variables for the latent variables are defined in the lines containing the 1*. Thus, in the solution for this model, neither the observed nor the latent variables are standardized. However, the keyword SC on the LISREL Output line will produce a standardized solution as a by-product (see Section 5.9). See Section 6.9 for general rules on how to define units of measurement for latent variables.

5.2 Classification of Variables

LISREL 8 classifies the observed and latent variables as follows.

- Eta(η)-variables: Eta1, Eta2.
  These are the dependent latent variables. All latent variables appearing on the left side of the equal sign in the relationships are Eta-variables. In the path diagram, they are recognized as those variables in circles or ellipses that have one-way (unidirectional) arrows pointing to them.

- Ksi(ξ)-variables: Ksi1, Ksi2, Ksi3.
  These are the remaining latent variables in the model. In the path diagram they are recognized as those variables in circles or ellipses that do not have one-way (unidirectional) arrows pointing to them.

- Y-variables: Y1, Y2, Y3, Y4.
  These are the observed variables which depend on Eta-variables.

- X-variables: X1, X2, X3, X4, X5, X6, X7.
  These are the observed variables which depend on Ksi-variables.

- Zeta(ζ)-variables: Zeta1, Zeta2.
  These are the error terms in the structural equations, i.e., the error terms on Eta1 and Eta2.

- Epsilon(ϵ)-variables: Epsilon1, Epsilon2, Epsilon3, Epsilon4.
  These are the measurement errors in the Y-variables. In the path diagram they are represented by one-way (unidirectional) arrows on the right side.

- Delta(δ)-variables: Delta1, Delta2, Delta3, Delta4, Delta5, Delta6, Delta7.
  These are the measurement errors in the X-variables. In the path diagram they are represented by one-way (unidirectional) arrows on the left side.

5.3 Parameter Matrices

Every one-way (unidirectional) arrow in the path diagram represents a parameter or coefficient. Depending on where the arrow is coming from or going to, these parameters have different names which correspond to Greek characters. In the following the mathematical Greek notation is given in parenthesis.

A path from an Eta-variable to another Eta-variable is called a BETA(β)-parameter, a path from a Ksi-variable to an Eta-variable is called a GAMMA(γ)-parameter, a path from a Eta-variable to a Y-variable is called a LAMBDA-Y($\lambda^{(y)}$)-parameter, or LY-parameter, for short, and a path from a Ksi-variable to a X-variable is called a LAMBDA-X($\lambda^{(x)}$)-parameter, or LX-parameter, for short.

Each parameter has two subscripts, the first being the index of the variable *to* which the path is going and the second being the index of the variable *from* which the path is coming. Thus, BETA(2,1) (β_{21}) is the parameter associated with the path from Eta1 (η_1) to Eta2 (η_2) and GAMMA(2,1) (γ_{21}) is the parameter associated with the path from Ksi1 (ξ_1) to Eta2 (η_2). In general, BETA(I,J) (β_{ij}) correspond to the path from Eta-j (η_j) to Eta-i (η_i), and GAMMA(K,L) (γ_{kl}) corresponds to the path from Ksi-l (ξ_l) to Eta-k (η_k). Similarly, LY(2,1) ($\lambda^{(y)}_{21}$) represents the path from Eta1 (η_1) to Y2 and LX(3,1) ($\lambda^{(x)}_{31}$) represents the path from Ksi1 (ξ_1) to X3. In general, LY(I,J) ($\lambda^{(y)}_{ij}$) represents the path from Eta-j (η_j) to Y-i and LX(K,L) ($\lambda^{(x)}_{kl}$) represents the path from Ksi-l(ξ_l) to X-k.

The BETA parameters may be collected in a matrix

$$\text{BETA} = \begin{pmatrix} 0 & \text{BETA(1,2)} \\ \text{BETA(2,1)} & 0 \end{pmatrix}$$

Similarly, the GAMMA parameters may be collected in a matrix

$$\text{GAMMA} = \begin{pmatrix} \text{GAMMA(1,1)} & \text{GAMMA(1,2)} & 0 \\ \text{GAMMA(2,1)} & 0 & \text{GAMMA(2,3)} \end{pmatrix}$$

The LY and LX parameters may also be collected in matrices

$$\text{LAMBDA-Y} = \begin{pmatrix} 1 & 0 \\ \text{LY(2,1)} & 0 \\ 0 & 1 \\ 0 & \text{LY(4,2)} \end{pmatrix}$$

$$\text{LAMBDA-X} = \begin{pmatrix} 1 & 0 & 0 \\ \text{LX(2,1)} & 0 & 0 \\ \text{LX(3,1)} & \text{LX(3,2)} & 0 \\ 0 & 1 & 0 \\ 0 & \text{LX(5,2)} & 0 \\ 0 & 0 & 1 \\ 0 & 0 & \text{LX(7,3)} \end{pmatrix}$$

Zero elements in these matrices correspond to non-existent paths in the path diagram. For example, since Eta1 (η_1) does not depend on Ksi3 (ξ_3), element GAMMA(1,3) (γ_{13}) should be zero. The ones in the matrices correspond to fixed ones according to the specification of the relationships in the input file. Note that the two indices i and j of a matrix element, which were previously defined to be the index of the "to-variable" and the "from-variable," respectively, now correspond to the row and column where the element is located within the matrix.

Using the Greek notation system, these matrices are written:

$$\mathbf{B} = \begin{pmatrix} 0 & \beta_{12} \\ \beta_{21} & 0 \end{pmatrix} \qquad \boldsymbol{\Gamma} = \begin{pmatrix} \gamma_{11} & \gamma_{12} & 0 \\ \gamma_{21} & 0 & \gamma_{23} \end{pmatrix}$$

$$\boldsymbol{\Lambda}_y = \begin{pmatrix} 1 & 0 \\ \lambda_{21}^{(y)} & 0 \\ 0 & 1 \\ 0 & \lambda_{42}^{(y)} \end{pmatrix} \qquad \boldsymbol{\Lambda}_x = \begin{pmatrix} 1 & 0 & 0 \\ \lambda_{21}^{(x)} & 0 & 0 \\ \lambda_{31}^{(x)} & \lambda_{32}^{(x)} & 0 \\ 0 & 1 & 0 \\ 0 & \lambda_{52}^{(x)} & 0 \\ 0 & 0 & 1 \\ 0 & 0 & \lambda_{73}^{(x)} \end{pmatrix}$$

It is also possible to specify the path diagram in Greek notation and include most of the parameters as coefficients attached to arrows. Such a path diagram is shown in Figure 5.2.

There are five additional parameter matrices, namely the covariance matrices of the Ksi-, Zeta-, Epsilon-, and Delta-variables and the covariance matrix between Delta- and Epsilon-variables. These are called

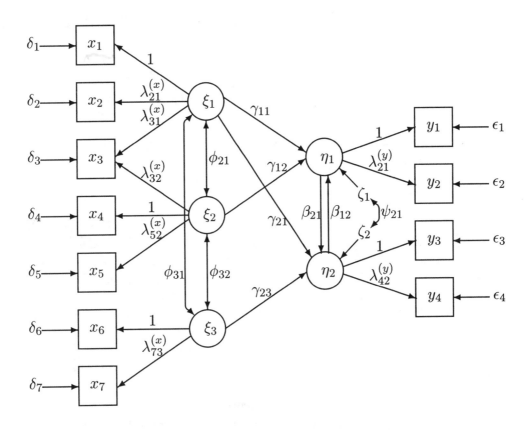

Figure 5.2   Path Diagram in Greek Notation

PHI(Φ), PSI(Ψ), THETA-EPS(Θ_ϵ), THETA-DELTA(Θ_δ), and THETA-DELTA-EPS($\Theta_{\delta\epsilon}$), respectively. The elements of PHI are the variances and covariances of the latent independent variables. The diagonal elements of PSI, THETA-EPS, and THETA-DELTA are the variances of the various error terms, i.e., the variances of the Zeta-, Epsilon-, and Delta-variables, respectively. Off-diagonal elements in these matrices represent error covariances and correspond to two-way arrows in the path diagram. The elements in THETA-DELTA-EPS represent covariances between measurement errors in X- and Y-variables, i.e., between Delta- and Epsilon-variables. In the hypothetical model, the only error covariance is PSI(2,1), the covariance between Zeta1 and Zeta2, the error terms on Eta1 and Eta2.

Using the Greek notational system, the first four of these matrices are:

$$\Phi = \begin{pmatrix} \phi_{11} & & \\ \phi_{21} & \phi_{22} & \\ \phi_{31} & \phi_{32} & \phi_{33} \end{pmatrix} \qquad \Psi = \begin{pmatrix} \psi_{11} & \\ \psi_{21} & \psi_{22} \end{pmatrix}$$

$$\Theta_\epsilon = \mathrm{diag}(\theta_{11}^{(\epsilon)}, \theta_{22}^{(\epsilon)}, \ldots, \theta_{44}^{(\epsilon)}) \qquad \Theta_\delta = \mathrm{diag}(\theta_{11}^{(\delta)}, \theta_{22}^{(\delta)}, \ldots, \theta_{77}^{(\delta)})$$

5.4 Parameter Specifications

LISREL 8 orders the parameter matrices as LAMBDA-Y, LAMBDA-X, BETA, GAMMA, PHI, PSI, THETA-EPS, THETA-DELTA-EPS, and THETA-DELTA, and the elements in these matrices row-wise. The parameters to be estimated are then labeled 1, 2, 3, etc. In the output file there is a section called **PARAMETER SPECIFICATIONS** where one can see how LISREL 8 has classified the variables in the model and which parameters are to be estimated.

The tables of parameter specifications consist of integer matrices corresponding to the parameter matrices. In each matrix an element is an integer equal to the index of the corresponding parameter in the sequence of independent parameters. Elements corresponding to fixed parameters are zero and elements constrained to be equal have the same index value. Parameter matrices which are entirely zero, i.e., which contain no parameters to be estimated, are omitted. In this example, THETA-DELTA-EPS is omitted because there are no covariances specified between Delta- and Epsilon-variables. For parameter matrices which are diagonal, only the diagonal elements are listed. In this example, THETA-EPS and THETA-DELTA are diagonal.

The parameter specifications for the hypothetical model look as follows:

```
PARAMETER SPECIFICATIONS
          LAMBDA-Y
          Eta1        Eta2
Y1           0           0
Y2           1           0
Y3           0           0
Y4           0           2
```

```
             LAMBDA-X
             Ksi1        Ksi2        Ksi3
     X1        0           0           0
     X2        3           0           0
     X3        4           5           0
     X4        0           0           0
     X5        0           6           0
     X6        0           0           0
     X7        0           0           7

             BETA
             Eta1        Eta2
   Eta1        0           8
   Eta2        9           0

             GAMMA
             Ksi1        Ksi2        Ksi3
   Eta1       10          11           0
   Eta2       12           0          13

             PHI
             Ksi1        Ksi2        Ksi3
   Ksi1       14
   Ksi2       15          16
   Ksi3       17          18          19

             PSI
             Eta1        Eta2
   Eta1       20
   Eta2       21          22

             THETA-EPS
              Y1          Y2          Y3          Y4
              23          24          25          26

             THETA-DELTA
              X1          X2          X3          X4          X5          X6          X7
              27          28          29          30          31          32          33
```

From this we can see that LISREL 8 has interpreted the relationships in the model correctly, that the parameter BE(2,1), say, is the ninth parameter, and that there are 33 parameters to be estimated altogether.

5.5   LISREL Estimates

In addition to the usual parameter estimates, LISREL 8 gives standard errors of estimates, and t-values and squared multiple correlations for each equation in the model. The estimated joint covariance matrix is also given. The information is the same as in the SIMPLIS output, but the parameter estimates are given in terms of parameter matrices instead of as equations. Zero elements in the parameter matrices are indicated by - -. The parameter matrices appear in a section of the output file called **LISREL ESTIMATES**. They are called LISREL estimates regardless of what method of estimation was used to produce them, but the method used is specified in parentheses after the header **LISREL ESTIMATES**. For our example, the parameter estimates in LISREL output format are:

```
         LAMBDA-Y
           Eta1        Eta2
  Y1       1.00         - -

  Y2        .92         - -
          (.04)
          25.99

  Y3        - -        1.00

  Y4        - -        1.14
                      (.03)
                      38.74


         LAMBDA-X
           Ksi1        Ksi2        Ksi3
  X1       1.00         - -         - -

  X2       1.29         - -         - -
          (.10)
          12.33

  X3        .92        1.09         - -
          (.12)       (.12)
           7.46        9.32

  X4        - -        1.00         - -

  X5        - -        1.08         - -
                      (.08)
                      12.80

  X6        - -         - -        1.00

  X7        - -         - -        1.44
                                  (.09)
                                  15.48
```

```
            BETA
            Eta1        Eta2
Eta1        - -         .54
                        (.06)
                        9.53


Eta2        .94         - -
            (.18)
            5.25



            GAMMA
            Ksi1        Ksi2        Ksi3
Eta1        .21         .50         - -
            (.15)       (.15)
            1.39        3.35


Eta2        -1.22       - -         1.00
            (.12)                   (.15)
            -10.05                  6.57

            COVARIANCE MATRIX OF ETA AND KSI
            Eta1        Eta2        Ksi1        Ksi2        Ksi3
Eta1        2.96
Eta2        3.12        4.72
Ksi1        .48         -.22        .97
Ksi2        .55         -.31        .79         1.12
Ksi3        .93         1.40        .52         .13         1.18

            PHI
            Ksi1        Ksi2        Ksi3
Ksi1        .97
            (.19)
            5.17


Ksi2        .79         1.12
            (.15)       (.20)
            5.21        5.72


Ksi3        .52         .13         1.18
            (.13)       (.13)       (.20)
            3.93        1.06        5.78


            PSI
            Eta1        Eta2
Eta1        .49
            (.13)
            3.83


Eta2        -.07        .13
            (.17)       (.08)
            -.41        1.70
```

```
SQUARED MULTIPLE CORRELATIONS FOR STRUCTURAL EQUATIONS
    Eta1        Eta2
    .84         .97

THETA-EPS
      Y1          Y2          Y3          Y4
     .25         .12         .14         .20
   (.05)       (.04)       (.04)       (.05)
    4.66        3.24        3.32        3.60


SQUARED MULTIPLE CORRELATIONS FOR Y - VARIABLES
      Y1          Y2          Y3          Y4
     .92         .95         .97         .97

THETA-DELTA
      X1          X2          X3          X4          X5          X6          X7
     .39         .34         .06         .26         .44         .25         .26
   (.06)       (.07)       (.05)       (.05)       (.07)       (.05)       (.09)
    6.09        5.00        1.23        5.14        5.88        4.60        2.83


SQUARED MULTIPLE CORRELATIONS FOR X - VARIABLES
      X1          X2          X3          X4          X5          X6          X7
     .71         .83         .98         .81         .75         .83         .90
```

It may be instructive to run the same input file without the line

```
LISREL Output: RS MI SC EF WP
```

to establish the fact that the two output files give the same information about the LISREL solution. The information is just displayed in different ways. As in the SIMPLIS output, the standard errors and the t-values are given below the parameter estimates.

5.6 Goodness-of-Fit Statistics

LISREL 8 gives many measures of the goodness-of-fit of the whole model. All fit measures are functions of chi-square, the fit measure appearing on the first line. These fit measures were defined and discussed in Chapter 4. The fit measures appear in a section of the output called **GOODNESS OF FIT STATISTICS**. For the hypothetical model, these measures are listed in the output file as:

```
                      GOODNESS OF FIT STATISTICS

         CHI-SQUARE WITH 33 DEGREES OF FREEDOM = 29.10 (P = 0.66)
              ESTIMATED NON-CENTRALITY PARAMETER (NCP) = 0.0
          90 PERCENT CONFIDENCE INTERVAL FOR NCP = (0.0 ; 12.26)


                  MINIMUM FIT FUNCTION VALUE = 0.29
            POPULATION DISCREPANCY FUNCTION VALUE (F0) = 0.0
          90 PERCENT CONFIDENCE INTERVAL FOR F0 = (0.0 ; 0.12)
         ROOT MEAN SQUARE ERROR OF APPROXIMATION (RMSEA) = 0.0
        90 PERCENT CONFIDENCE INTERVAL FOR RMSEA = (0.0 ; 0.061)
          P-VALUE FOR TEST OF CLOSE FIT (RMSEA ¡ 0.05) = 0.90


              EXPECTED CROSS-VALIDATION INDEX (ECVI) = 0.96
         90 PERCENT CONFIDENCE INTERVAL FOR ECVI = (1.00 ; 1.12)
                   ECVI FOR SATURATED MODEL = 1.33

    CHI-SQUARE FOR INDEPENDENCE MODEL WITH 55 DEGREES OF FREEDOM = 1441.95
                    INDEPENDENCE AIC = 1463.95
                         MODEL AIC = 95.10
                      SATURATED AIC = 132.00
                   INDEPENDENCE CAIC = 1503.61
                        MODEL CAIC = 214.07
                     SATURATED CAIC = 369.94


              ROOT MEAN SQUARE RESIDUAL (RMR) = 0.065
                    STANDARDIZED RMR = 0.027
                GOODNESS OF FIT INDEX (GFI) = 0.95
          ADJUSTED GOODNESS OF FIT INDEX (AGFI) = 0.91
         PARSIMONY GOODNESS OF FIT INDEX (PGFI) = 0.48


                 NORMED FIT INDEX (NFI) = 0.98
               NON-NORMED FIT INDEX (NNFI) = 1.00
            PARSIMONY NORMED FIT INDEX (PNFI) = 0.59
               COMPARATIVE FIT INDEX (CFI) = 1.00
               INCREMENTAL FIT INDEX (IFI) = 1.00
                RELATIVE FIT INDEX (RFI) = 0.97


                       CRITICAL N (CN) = 187.35
```

These statistics illustrate the situation of a small random sample from a population in which the specified model holds exactly. In situations of large samples of real data or of random samples from populations in which the specified model does not hold, these statistics may look quite different.

5.7 Fitted and Standardized Residuals

A residual is an observed minus a fitted covariance (variance). A standardized residual is a residual divided by its estimated standard error. There are such residuals for every pair of observed variables. Fitted residuals depend on the unit of measurement of the observed variables. If the variances of the variables vary considerably from one variable to another, it is rather difficult to know whether a fitted residual should be considered large or small. Standardized residuals, on the other hand, are independent of the units of measurement of the variables and provides a "statistical" metric for judging the size of a residual.

The fitted and standardized residuals for the hypothetical model are typical for the situation when the data fits the model well. In the output file they look like this.

```
                FITTED RESIDUALS
          Y1     Y2     Y3     Y4     X1     X2     X3     X4     X5     X6     X7
   Y1    .00
   Y2    .00    .00
   Y3    .08    .01    .00
   Y4    .00   -.06    .00    .00
   X1   -.15   -.07   -.14   -.22    .00
   X2   -.06    .02   -.04   -.02    .01    .00
   X3   -.04    .05    .05    .02   -.01    .01    .00
   X4   -.09   -.05   -.13   -.18    .00    .03    .01    .00
   X5   -.10   -.01   -.03   -.04   -.01   -.03   -.01   -.02    .00
   X6    .12    .10    .01    .12   -.05    .02    .03   -.06   -.04    .00
   X7   -.08   -.08   -.09    .02   -.07   -.07    .01   -.06   -.04    .00    .00

                STANDARDIZED RESIDUALS
          Y1     Y2     Y3     Y4     X1     X2     X3     X4     X5     X6     X7
   Y1    .00
   Y2    .00    .00
   Y3   1.66    .21    .00
   Y4   -.04  -1.66    .00    .00
   X1  -1.50   -.83  -1.14  -1.59    .00
   X2   -.71    .27   -.36   -.14    .58    .00
   X3   -.56   1.27    .93    .35   -.45    .35    .00
   X4  -1.02   -.77  -1.29  -1.63    .00    .60    .53    .00
   X5   -.86   -.12   -.20   -.27   -.20   -.48   -.37  -1.08    .00
   X6   1.64   1.74    .20   1.45   -.76    .26    .35   -.98   -.49    .00
   X7  -1.11  -1.58  -1.55    .23   -.84  -1.01    .25   -.79   -.46    .00    .00
```

All the standardized residuals may be examined collectively in a Q-plot. A good model is characterized by points falling approximately on a straight line. Deviations from this pattern are indicative of specification errors in the model, non-normality in the variables or of nonlinear relationships among the variables. In particular, standardized residuals which appear as outliers in the Q-plot are indicative of a specification error in the model.

The Q-plot of the standardized residuals for the hypothetical model is given on page 148.

5.8 Modification Indices

A modification index (Sörbom 1989) is given for each fixed and constrained parameter in the model. Each such modification index measures how much chi-square is expected to decrease if this particular parameter is set free and the model is reestimated. Thus, the modification index is approximately equal to the difference in chi-square between two models in which one parameter is fixed or constrained in one model and free in the other model, all other parameters estimated in both models. The largest modification index tells which parameter to set free to improve the fit maximally.

Associated with each modification index is an estimated change of the parameter. This measures how much the parameter is expected to change, in the positive or negative direction, if set free. In addition, the expected change is given in the metric when the latent variables are standardized and in the metric when both the observed and the latent variables are standardized, as suggested by Kaplan (1989). For the hypothetical model, the modification indices and the expected changes are:

```
MODIFICATION INDICES AND EXPECTED CHANGE

     MODIFICATION INDICES FOR LAMBDA-Y
           Eta1       Eta2
  Y1        - -        .99
  Y2        - -        .99
  Y3       2.11        - -
  Y4       2.11        - -
```

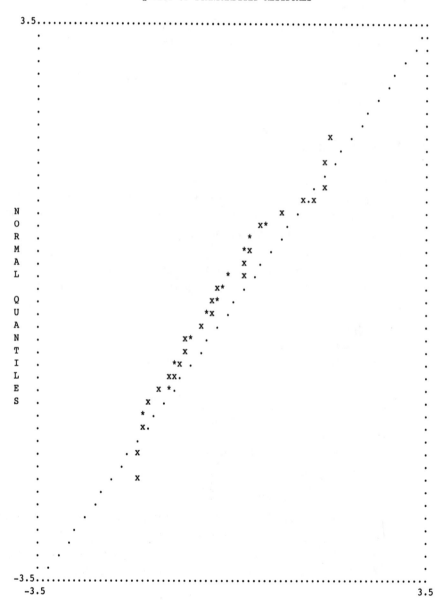

Q-PLOT OF STANDARDIZED RESIDUALS

```
EXPECTED CHANGE FOR LAMBDA-Y
        Eta1     Eta2
Y1      - -       .06
Y2      - -      -.05
Y3      .09      - -
Y4     -.10      - -

STANDARDIZED EXPECTED CHANGE FOR LAMBDA-Y
        Eta1     Eta2
Y1      - -       .13
Y2      - -      -.12
Y3      .15      - -
Y4     -.17      - -

COMPLETELY STANDARDIZED EXPECTED CHANGE FOR LAMBDA-Y
        Eta1     Eta2
Y1      - -       .07
Y2      - -      -.07
Y3      .07      - -
Y4     -.07      - -

MODIFICATION INDICES FOR LAMBDA-X
        Ksi1     Ksi2     Ksi3
X1      - -       .11     1.05
X2      - -       .11      .34
X3      - -      - -      3.38
X4      .30      - -      1.37
X5      .30      - -       .24
X6      .21      .00      - -
X7      .21      .00      - -

EXPECTED CHANGE FOR LAMBDA-X
        Ksi1     Ksi2     Ksi3
X1      - -      -.04     -.08
X2      - -       .06     -.05
X3      - -      - -       .15
X4      .08      - -      -.08
X5     -.09      - -      -.04
X6      .03      .00      - -
X7     -.05      .00      - -

STANDARDIZED EXPECTED CHANGE FOR LAMBDA-X
        Ksi1     Ksi2     Ksi3
X1      - -      -.05     -.09
X2      - -       .06     -.06
X3      - -      - -       .16
X4      .08      - -      -.08
X5     -.09      - -      -.04
X6      .03      .00      - -
X7     -.05      .00      - -
```

```
       COMPLETELY STANDARDIZED EXPECTED CHANGE FOR LAMBDA-X
              Ksi1         Ksi2         Ksi3
     X1        - -         -.04         -.08
     X2        - -          .04         -.04
     X3        - -          - -          .08
     X4        .07          - -         -.07
     X5       -.07          - -         -.03
     X6        .03          .00          - -
     X7       -.03          .00          - -


NO NON-ZERO MODIFICATION INDICES FOR BETA
NO NON-ZERO MODIFICATION INDICES FOR GAMMA
NO NON-ZERO MODIFICATION INDICES FOR PHI
NO NON-ZERO MODIFICATION INDICES FOR PSI


       MODIFICATION INDICES FOR THETA-EPS
              Y1           Y2           Y3           Y4
     Y1        - -
     Y2        - -          - -
     Y3        .53          .00          - -
     Y4        .55          .00          - -          - -


       EXPECTED CHANGE FOR THETA-EPS
              Y1           Y2           Y3           Y4
     Y1        - -
     Y2        - -          - -
     Y3        .02          .00          - -
     Y4       -.03          .00          - -          - -


       COMPLETELY STANDARDIZED EXPECTED CHANGE FOR THETA-EPS
              Y1           Y2           Y3           Y4
     Y1        - -
     Y2        - -          - -
     Y3        .01          .00          - -
     Y4       -.01          .00          - -          - -


       MODIFICATION INDICES FOR THETA-DELTA-EPS
              Y1           Y2           Y3           Y4
     X1        .13          .50         2.04         4.21
     X2        .24          .10          .35          .62
     X3        .00          .06          .01          .05
     X4        .91          .03          .04          .37
     X5       1.03          .10          .11          .71
     X6        .95         1.12         2.40          .17
     X7        .29         1.70          .02          .91


       EXPECTED CHANGE FOR THETA-DELTA-EPS
              Y1           Y2           Y3           Y4
     X1       -.01          .02          .05         -.08
     X2       -.02          .01         -.02          .03
     X3        .00         -.01          .00          .01
     X4        .03          .00         -.01         -.02
     X5       -.04          .01         -.01          .03
```

X6	.03	.03	-.05	.01
X7	-.02	-.05	.01	.05

COMPLETELY STANDARDIZED EXPECTED CHANGE FOR THETA-DELTA-EPS

	Y1	Y2	Y3	Y4
X1	-.01	.01	.02	-.03
X2	-.01	.00	-.01	.01
X3	.00	.00	.00	.00
X4	.02	.00	.00	-.01
X5	-.02	.01	.00	.01
X6	.02	.02	-.02	.00
X7	-.01	-.02	.00	.01

MODIFICATION INDICES FOR THETA-DELTA

	X1	X2	X3	X4	X5	X6	X7
X1	- -						
X2	.34	- -					
X3	.57	.02	- -				
X4	.00	.09	.18	- -			
X5	.14	.29	2.01	1.16	- -		
X6	.46	.81	.08	.59	.16	- -	
X7	.60	1.93	.52	.50	.05	- -	- -

EXPECTED CHANGE FOR THETA-DELTA

	X1	X2	X3	X4	X5	X6	X7
X1	- -						
X2	.03	- -					
X3	-.03	.01	- -				
X4	.00	.01	-.03	- -			
X5	.02	-.03	.11	-.08	- -		
X6	-.03	.04	-.01	-.03	-.02	- -	
X7	.04	-.08	.03	.03	-.01	- -	- -

COMPLETELY STANDARDIZED EXPECTED CHANGE FOR THETA-DELTA

	X1	X2	X3	X4	X5	X6	X7
X1	- -						
X2	.02	- -					
X3	-.01	.00	- -				
X4	.00	.01	-.01	- -			
X5	.01	-.01	.04	-.05	- -		
X6	-.02	.02	.00	-.02	-.01	- -	
X7	.02	-.03	.01	.02	-.01	- -	- -

MAXIMUM MODIFICATION INDEX IS   4.21 FOR ELEMENT (1, 4) OF THETA DELTA-EPSILON

5.9 Standardized Solutions

The latent variables in the LISREL solution are generally standardized. In this example, however, neither the latent nor the observed variables are standardized. Nevertheless, it is possible to obtain standardized solutions as a by-product after the original solution has been obtained. There are two kinds of standardized solutions: SS (Standardized Solution), in which the latent variables are scaled to have variances equal to one and the *observed variables are still in the original metric* and SC (for Standardized Completely), in which both observed and latent variables are standardized.

These standardized solutions can only be computed after the original solution has been estimated. To obtain these solutions, one must first obtain diagonal matrices of estimated standard deviations of the latent variables (SS) and the observed variables (SC).

The standard deviations of the latent variables are obtained from the joint covariance matrix of the Eta- and Ksi-variables which is given in the section **LISREL ESTIMATES**:

```
        COVARIANCE MATRIX OF ETA AND KSI
                Eta1        Eta2        Ksi1        Ksi2        Ksi3
        Eta1    2.96
        Eta2    3.12        4.72
        Ksi1     .48        -.22         .97
        Ksi2     .55        -.31         .79        1.12
        Ksi3     .93        1.40         .52         .13        1.18
```

The standard deviations of the latent variables are the square roots of the diagonal elements of this matrix.

The standard deviations of the observed variables are the square roots of the diagonal elements of the fitted covariance matrix. For the hypothetical model this is:

```
        FITTED COVARIANCE MATRIX
          Y1     Y2     Y3     Y4     X1     X2     X3     X4     X5     X6     X7
    Y1  3.20
    Y2  2.72   2.63
    Y3  3.12   2.87   4.85
    Y4  3.55   3.27   5.37   6.31
    X1   .48    .44   -.22   -.25   1.36
    X2   .62    .57   -.28   -.32   1.26   1.96
    X3  1.05    .96   -.54   -.61   1.76   2.27   3.80
    X4   .55    .51   -.31   -.35    .79   1.02   1.95   1.38
    X5   .60    .55   -.34   -.38    .85   1.10   2.10   1.21   1.74
    X6   .93    .86   1.40   1.60    .52    .68    .63    .13    .14   1.42
    X7  1.34   1.23   2.01   2.29    .75    .97    .90    .19    .21   1.69   2.68
```

The standardized solutions SS and SC are computed by applying these standard deviations or their reciprocals as scale factors in the rows and columns of the estimated parameter matrices of the LISREL solution. For formulas, see Jöreskog & Sörbom (1989, pp. 38–39). Note that the standard deviations of the observed variables are obtained from the fitted covariance matrix rather than the observed covariance matrix. For the hypothetical model, the completely standardized solution (SC) is:

```
COMPLETELY STANDARDIZED SOLUTION
```

LAMBDA-Y

	Eta1	Eta2
Y1	.96	- -
Y2	.98	- -
Y3	- -	.99
Y4	- -	.98

LAMBDA-X

	Ksi1	Ksi2	Ksi3
X1	.85	- -	- -
X2	.91	- -	- -
X3	.47	.59	- -
X4	- -	.90	- -
X5	- -	.86	- -
X6	- -	- -	.91
X7	- -	- -	.95

BETA

	Eta1	Eta2
Eta1	- -	.68
Eta2	.74	- -

GAMMA

	Ksi1	Ksi2	Ksi3
Eta1	.12	.30	- -
Eta2	-.56	- -	.50

CORRELATION MATRIX OF ETA AND KSI

	Eta1	Eta2	Ksi1	Ksi2	Ksi3
Eta1	1.00				
Eta2	.83	1.00			
Ksi1	.28	-.10	1.00		
Ksi2	.30	-.14	.76	1.00	
Ksi3	.50	.60	.49	.12	1.00

PSI

	Eta1	Eta2
Eta1	.16	
Eta2	-.02	.03

THETA-EPS

Y1	Y2	Y3	Y4
.08	.05	.03	.03

```
THETA-DELTA
    X1          X2          X3          X4          X5          X6          X7
   .29         .17         .02         .19         .25         .17         .10
```

5.10   Direct, Indirect, and Total Effects (EF)

It can be seen in the path diagram of Figure 4.1 that there are both direct and indirect effects of Ksi1 on Eta2. For example, in addition to the direct effect GAMMA(2,1) of Ksi1 on Eta2, there is an indirect effect BETA(2,1)$\times$GAMMA(1,1) mediated by Eta1. And even though there is no direct effect of Ksi3 on Eta1, there is a similar indirect effect BETA(1,2)$\times$GAMMA(2,3) mediated by Eta2.

There are usually no direct effects of an Eta-variable on itself, i.e., all diagonal elements of the BETA matrix are zero. Nevertheless, there may be a total effect of an Eta-variable on itself. How can this be? This can only occur in non-recursive models and can best be understood by defining a cycle. A cycle is a causal chain going from one Eta-variable, passing over some other Eta-variables and returning to the original Eta-variable. The two Eta-variables in Figure 5.1 are shown in isolation in Figure 5.3.

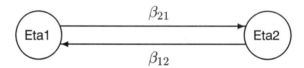

Figure 5.3   Reciprocal Causation Between Eta1 and Eta2

One cycle for Eta1 consists of one path to Eta2 and a return to Eta1. Writing β_{21} for BETA(2,1) and β_{12} for BETA(1,2), the effect of one cycle on Eta1 is $\beta_{21}\beta_{12}$. After two cycles the effect will be $\beta_{21}^2\beta_{12}^2$, after three cycles $\beta_{21}^3\beta_{12}^3$, etc. The total effect on Eta1 will be the sum of the infinite geometric series

$$\beta_{21}\beta_{12} + \beta_{21}^2\beta_{12}^2 + \beta_{21}^3\beta_{12}^3 \cdots \tag{5.1}$$

which is $\beta_{21}\beta_{12}/(1 - \beta_{21}\beta_{12})$ if $|\beta_{21}\beta_{12}| < 1$.

A necessary and sufficient condition for convergence of the series in (5.1) is that all the eigenvalues of the BETA matrix are within the unit circle. In general the eigenvalues of a BETA matrix are complex numbers somewhat difficult to compute. However, a sufficient condition for convergence is that the largest eigenvalue of $\mathbf{BB'}$ is less than one. This is very easy to verify. The program prints the largest eigenvalue of $\mathbf{BB'}$ under the name of **STABILITY INDEX**. For a fuller discussion of indirect and total effects, see Bollen (1987, 1989a).

The indirect and total effects will be computed and printed if requested by the EF keyword on the LISREL Output line. Standard errors of indirect and total effects can also be computed (Sobel, 1982). In LISREL 8, these are automatically included in the LISREL Output, together with the corresponding t-values.

For the hypothetical model, some of the estimated indirect and total effects and their standard errors are:

```
TOTAL AND INDIRECT EFFECTS

    TOTAL EFFECTS OF KSI ON  ETA

              Ksi1      Ksi2      Ksi3
    Eta1      -.90      1.00      1.08
             (.43)     (.34)     (.22)
             -2.09      2.92      4.81

    Eta2     -2.06       .94      2.01
             (.52)     (.40)     (.27)
             -3.99      2.32      7.49

    INDIRECT EFFECTS OF KSI ON  ETA

              Ksi1      Ksi2      Ksi3
    Eta1     -1.11       .50      1.08
             (.34)     (.24)     (.22)
             -3.27      2.10      4.81

    Eta2      -.84       .94      1.01
             (.47)     (.40)     (.31)
             -1.80      2.32      3.27

    TOTAL EFFECTS OF ETA ON  ETA

              Eta1      Eta2
    Eta1      1.02      1.08
             (.41)     (.28)
              2.48      3.85
```

```
Eta2        1.89        1.02
           (.72)       (.41)
            2.64        2.48
```

LARGEST EIGENVALUE OF B*B' (STABILITY INDEX) IS   .879

INDIRECT EFFECTS OF ETA ON ETA

```
            Eta1        Eta2
Eta1        1.02         .55
           (.41)       (.25)
            2.48        2.21

Eta2         .95        1.02
           (.55)       (.41)
            1.74        2.48
```

TOTAL EFFECTS OF ETA ON Y

```
            Eta1        Eta2
Y1          2.02        1.08
           (.41)       (.28)
            4.91        3.85

Y2          1.86        1.00
           (.38)       (.26)
            4.83        3.86

Y3          1.89        2.02
           (.72)       (.41)
            2.64        4.91

Y4          2.15        2.30
           (.82)       (.47)
            2.63        4.87
```

INDIRECT EFFECTS OF ETA ON Y

```
            Eta1        Eta2
Y1          1.02        1.08
           (.41)       (.28)
            2.48        3.85

Y2           .94        1.00
           (.38)       (.26)
            2.47        3.86
```

```
Y3        1.89        1.02
         (.72)       (.41)
          2.64        2.48

Y4        2.15        1.16
         (.82)       (.47)
          2.63        2.47
```

TOTAL EFFECTS OF KSI ON   Y

	Ksi1	Ksi2	Ksi3
Y1	-.90	1.00	1.08
	(.43)	(.34)	(.22)
	-2.09	2.92	4.81
Y2	-.82	.92	.99
	(.39)	(.31)	(.21)
	-2.09	2.92	4.84
Y3	-2.06	.94	2.01
	(.52)	(.40)	(.27)
	-3.99	2.32	7.49
Y4	-2.35	1.07	2.29
	(.59)	(.46)	(.31)
	-3.99	2.32	7.48

Since there are no direct effects of KSI on Y, the indirect effects are equal to the total effects; hence only the total effects are given.

If SS or SC appears on the LISREL Output line, the corresponding standardized effects will also be obtained in the output file. These are the effects that would be obtained if the latent (SS) and the observed variables (SC) were standardized. For our example, the completely standardized effects are:

STANDARDIZED TOTAL AND INDIRECT EFFECTS

COMPLETELY STANDARDIZED TOTAL EFFECTS OF ETA ON   Y

	Eta1	Eta2
Y1	1.94	1.32
Y2	1.97	1.34
Y3	1.47	1.99
Y4	1.47	1.98

```
        COMPLETELY STANDARDIZED INDIRECT EFFECTS OF ETA ON   Y
                 Eta1        Eta2
        Y1        .98        1.32
        Y2        .99        1.34
        Y3       1.47        1.00
        Y4       1.47        1.00

        COMPLETELY STANDARDIZED TOTAL EFFECTS OF KSI ON   Y
                 Ksi1        Ksi2        Ksi3
        Y1       -.49         .59         .65
        Y2       -.50         .60         .66
        Y3       -.92         .45         .99
        Y4       -.92         .45         .99
```

5.11   Estimating the Standardized Solution Directly

It is instructive to estimate the standardized solution directly and compare the unstandardized and the standardized solutions. To estimate the completely standardized solution, one must analyze a correlation matrix and omit the 1* in the relationships which specify that the latent variables are to be scaled in the metric of observed reference variables. The following input file will estimate the completely standardized solution directly (see **EX17B.SPL**).

```
Example 17B: Hypothetical Model - Standardized Solution
Observed Variables: Y1-Y4 X1-X7
Correlation Matrix from File EX17.COR
Sample Size: 100
Latent Variables: Eta1 Eta2 Ksi1-Ksi3
Relationships
    Eta1 = Eta2 Ksi1 Ksi2
    Eta2 = Eta1 Ksi1 Ksi3
    Let the Errors of Eta1 and Eta2 Correlate

    Y1 - Y2 = Eta1
    Y3 - Y4 = Eta2

    X1 - X3 = Ksi1
    X3 - X5 = Ksi2
    X6 - X7 = Ksi3

LISREL Output: RS MI SC EF WP
End of Problem
```

Verify that the following quantities are the same in the output files for **EX17A.SPL** and **EX17B.SPL**.

- ☐ All goodness-of-fit statistics
- ☐ Standardized residuals and Q-plot
- ☐ Modification indices
- ☐ Completely standardized expected changes
- ☐ Completely standardized solution
- ☐ Completely standardized total effects

6 SIMPLIS REFERENCE

The examples of the previous chapters illustrate most of the features and options of LISREL 8 using the SIMPLIS command language. General syntax rules for the SIMPLIS command language and all the options are defined in general terms in this chapter.

6.1 Input File

An input file must contain ASCII characters only. The input file may contain title lines, command lines, data, and comments. Upper case and lower case characters may be used interchangeably except in names of variables, see Section 6.3. The exclamation mark (!) or the slash-asterisk combination (/*) may be used to indicate that everything that follows on this line is to be regarded as comments.

The term *physical line* is used here in the sense of an input record which ends with a RETURN and/or LINE FEED character. A SIMPLIS *command line* ends either with a RETURN and/or LINE FEED character or a semicolon (;). By using a semicolon to end a line, several lines can be put on the same physical line. Thus, for example, the physical line

```
Covariance Matrix from File EX10.COV;Sample Size = 865
```

consists of the two SIMPLIS command lines:

```
Covariance Matrix from File EX10.COV
Sample Size = 865
```

An input file in the SIMPLIS command language consists of a number of header lines, each followed by the type of information that the header indicates. A typical input file follows. Here, optional header lines and optional information on the header lines are enclosed within square brackets.

161

```
[Title]
- - - - - - - -
- - - - - - - -
Observed Variables [from File filename]
- - - - - - - -
- - - - - - - -
Covariance Matrix [from File filename]
- - - - - - - -
- - - - - - - -
Sample Size

Relationships
- - - - - - - -
- - - - - - - -
[Method = Unweighted Least-Squares]
[Number of Decimals = 3]
[Iterations = 100]
[Options: [RS WP AD=OFF]]

[LISREL Output]

[Path Diagram]

[End of Problem]
```

Although this example is somewhat typical, it does not include all header lines or all information that can be put on the lines. All header lines and all information on them are presented in the following sections. This material is presented in the order it normally arises in the preparation of the input file.

6.2   Title

The first line for each problem may be a title line containing any information used as a heading for the problem. One may choose not to have a title line or use only a single title line. However, any number of title lines may be used to describe the model and the data. The program will read title lines until it finds one of the following:

- A physical line beginning with the words Observed Variables or Labels, which is the first command line in a SIMPLIS input file.
- A physical line whose first two non-blank characters are DA, Da, dA, or da, which is the first command line in a LISREL input file.

Therefore, one must not use title lines that begin with these characters. To avoid this conflict, begin every title line with !. Then anything can appear on the title lines.

Title lines are optional but strongly recommended. Only title lines may appear before the first genuine command line.

Important note: Although title lines are optional in single-sample problems, at least one title line is necessary for each group except the first in multi-sample problems, and the first title line for each group must begin with the word Group ***(see Chapter 2)***.

6.3 Observed Variables or Labels

After the title lines, if any, a header line must follow with the words Observed Variables or Labels. The reason for the qualification Observed is that LISREL deals also with variables that are unobserved, the so-called latent variables, whose names must also be defined. In the SIMPLIS command language, the words Labels and Observed Variables are synonymous.

After the line with the words Observed Variables or Labels, enter a space (space character) or a colon and a space and follow with a list of names (labels) of the observed variables in the data. The names can begin on the header line itself or on the next line or they can be read from a file, in which case the command line is:

Observed Variables from File = *filename*

or

Labels from File = *filename*

The labels determine both the *number* of variables and the *order* of the variables. Each label may consist of any number of characters, but only the first eight characters will be retained and printed by the program.

The labels are entered in free format. Spaces, commas, and return characters (carriage returns or line feeds) are used as delimiters. Therefore, spaces and commas cannot be used within a label unless the label is enclosed in single quotes. The same holds for - (dash or minus sign), which has a special meaning, see below. Do not break a line in the middle of a label. The following sets of labels are all equivalent:

```
'VIS PERC' CUBES LOZENGES 'PAR COMP' 'SEN COMP' WORDMEAN

'VIS PERC'  CUBES LOZENGES 'PAR COMP', 'SEN COMP',
WORDMEAN

'VIS PERCEPTION' CUBES LOZENGES 'PAR COMP' 'SEN COMPLETION'
WORDMEANING
```

Labels are case sensitive; upper case or lower case can be used without restriction but one must use the same name to refer to the same variable each time. In this book we have used upper case words to refer to names of observed variables and capitalized words to refer to latent variables.

In the printed output, the labels will be right-adjusted within a field of eight characters. Blank spaces will be filled in before a label if it contains fewer than eight characters. But you need not take these spaces into account when referring to labels.

It is recommended that each variable be given a unique name. However, if you do not want to name all the variables, the following options are available.

If there are 123 variables, say, the line

```
VAR1 - VAR123
```

will automatically name the variables VAR1, VAR2, ..., VAR123. Even more simply, one can use the line

```
1 - 123
```

to label the variables 1, 2, ..., 123.

The line

```
YVAR1 - YVAR52, XVAR1 - XVAR71
```

will also label all 123 variables.

A line of the following form may also be used:

```
'VIS PERC' CUBES LOZENGES YVAR4 - YVAR52
HEIGHT WEIGHT AGE IQ XVAR5 -XVAR71
```

Here, some of the variables are given unique names; others are given names that begin with YVAR or XVAR.

The general rule is that if two labels end with integers m and n, with m less than n, one can use a - (dash or minus sign) to name all variables that end with consecutive integers from m through n. Whatever appears before the integer m will also appear before all the other integers.

One can also assign labels to end with consecutive characters instead of integers. For example, the line

```
VARa - VARk
```

will assign labels VARa, VARb, VARc, . . ., VARk, and the line

```
VARA - VARK
```

will assign labels VARA, VARB, VARC, . . ., VARK.

These "tricks" will work also with labels containing a - if the labels are enclosed within single quotes. Thus, the line

```
'Eta-1' - 'Eta-7'
```

will label variables Eta-1, Eta-2, . . ., Eta-7. Without the quotes, the line is disastrous.

6.4 Data

Data for LISREL 8 may be one of the following:

- ☐ Raw data
- ☐ Covariance matrix
- ☐ Covariance matrix and means
- ☐ Correlation matrix
- ☐ Correlation matrix and standard deviations
- ☐ Correlation matrix, standard deviations, and means

In addition to one of these, an asymptotic covariance matrix or a set of asymptotic variances may be required.

6.4.1 Raw Data

Although LISREL 8 can read raw data in free format directly from the input file, as Example 2 of Chapter 1 illustrates, it is strongly recommended that the raw data be stored in an external file to be read by LISREL 8. To read the raw data from a file, write

Raw Data from File *filename*

If the first line of the data file is a FORTRAN format (beginning with a left parenthesis and ending with a right parenthesis), the data will be read according to that format; otherwise, the data will be read in free format with spaces between entries, starting on a new line for each case.

LISREL 8 cannot deal with missing data or other problems in the raw data. If you have such problems, you should first use the program PRELIS (Jöreskog & Sörbom, 1988) to deal with all data problems and to compute an appropriate covariance or correlation matrix depending on the type of observed variables. If some or all of the variables are ordinal or highly non-normal, PRELIS can also estimate the asymptotic covariance matrix (or asymptotic variances) of the elements of the covariance or correlation matrix for use with the WLS (or DWLS)—see Section 6.14.4. The matrices produced by PRELIS can be read by LISREL 8 from an external file (see Sections 6.4.4 and 6.4.5).

6.4.2 Covariance Matrix or Correlation Matrix

The covariance or correlation matrix of the variables is a symmetric matrix, of which only the elements below and including the diagonal need to be given. If k labels have been read for the observed variables, the covariance or correlation matrix is of order $k \times k$ and contains $k(k+1)/2$ distinct elements. These elements are entered row-wise, each row beginning with the element in the first column of the matrix and ending with the diagonal element. The elements of the covariance or correlation matrix may be entered either in free format or in a user specified FORTRAN format. The easiest and most convenient way is to use a free format. In free format, commas, blanks and carriage returns are used as delimiters.

Example: Suppose, the covariance matrix is

$$
\begin{pmatrix}
38.60 & & & \\
13.63 & 16.96 & & \\
24.62 & 8.00 & 27.22 & \\
5.60 & 4.81 & 6.27 & 6.16
\end{pmatrix}
$$

The following ways of reading this matrix are all equivalent.

```
Covariance Matrix
38.60
13.63 16.96
24.62  8.00 27.22
 5.60  4.81  6.27 6.16

Covariance Matrix
38.6 13.63 16.96 24.62 8
27.22  5.6,4.81,6.27,6.16

Covariance Matrix
38.6 13.63 16.96 24.62 8 27.22 5.6 4.81 6.27 6.16
```

Any number of spaces may be used between the numbers.

The elements of the covariance matrix may also follow directly on the same line as the Covariance Matrix; thus

```
Covariance Matrix: 38.6 13.63 16.96 24.62 8
                   27.22  5.6,4.81,6.27,6.16
```

or

```
Covariance Matrix: 38.6 13.63 16.96 24.62 8 27.22 5.6 4.81 6.27 6.16
```

The colon is optional but there must be at least one space before the first entry of the covariance matrix.

The elements of the covariance or correlation matrix may also be entered according to a user specified F-format. To do so, the format must precede the elements on a separate line. Using the same example, one way to read formatted data is:

```
Covariance Matrix
(10F5.2)
 3860 1363 1696 2462  800 2722  560  481  627  616
```

The format (10F5.2) means that each element has a field of five positions and the last two digits are decimals, i.e., the program will insert a decimal point after the third position in each field. The number 10 in the format is the total number of elements in the matrix.

Note that elements of the lower half of the covariance matrix are entered as one long line. There must not be a carriage return or line feed within this line, only after the line. This means, for example, that if your matrix is of order 10, the total number of elements in the matrix is $1/2 \times 10 \times 11 = 55$ and if this matrix is entered with the format (55F5.2), the line will be $5 \times 55 = 275$ characters long and no carriage returns are allowed anywhere within this line. If your terminal or computer has wrap-around in column 80, this line will appear on the screen as three lines of 80 characters plus one line of 35 characters. For further details about FORTRAN formats, consult a FORTRAN manual.

One can also enter the matrix in a symmetric way by giving the word SYMMETRIC on the format line after the format:

```
Covariance Matrix
(4F5.2) SYMMETRIC
 3860
 1363 1696
 2462  800 2722
  560  481  627  616
```

Only the letter S in the word SYMMETRIC is significant, so the format line can also be written as:

```
(4F5.2) S
```

In this case there is a line for each row of the matrix, and each line ends with a carriage return after the diagonal element. The number 4 in the format corresponds to the order of the matrix or the total number of elements in the last row of the matrix.

One can also specify that the *full* matrix should be entered by giving the word FULL on the format line:

```
Covariance Matrix
(4F5.2) FULL
 3860 1363 2462   560
 1363 1696  800   481
 2462  800 2722   627
  560  481  627   616
```

Only the letter F in FULL is significant. This option is provided only because some programs produce covariance and correlation matrices in this redundant form.

Rather than putting the covariance or correlation matrix in the input file, one can put the matrix in an external file and tell LISREL 8 to read it from that file. This is particularly convenient if the matrix has been computed by PRELIS or some other program and already stored in a file. To read a correlation matrix, say, from an external file, write:

```
Correlation Matrix from File filename
```

If the first line of the external file is a FORTRAN format beginning with (and ending with), the matrix will be read according to that format; otherwise the matrix will be read in free format requiring spaces between entries. The rules for reading formatted and unformatted data as given above apply to the external file as well. See Examples 4, 5, 7, 8, and 9 of Chapter 1 on how to read covariance or correlation matrices from external files.

6.4.3 Means and Standard Deviations

The command line for reading means is

```
Means: 1.051 2.185 3.753
```

or
```
Means:
1.051 2.185 3.753
```

Similarly, the command line for reading standard deviations is

```
Standard Deviations: 0.984 1.107  .893
```

or
```
Standard Deviations
0.984 1.107  .893
```

The means and/or the standard deviations may also be read from an external file by adding from File *filename* to the command line. If the number of observed variables is k, there must be k means and standard deviations. The same rules for reading formatted and unformatted data as given in the previous section apply.

6.4.4 Asymptotic Covariance Matrix

An asymptotic covariance matrix is needed to obtain estimates by the weighted least-squares method (WLS). This matrix is computed by PRELIS and is stored in a file in binary form, see Jöreskog & Sörbom (1993a–b). To read it, put the line

```
Asymptotic Covariance Matrix from File filename
```

in the SIMPLIS input file.

6.4.5 Asymptotic Variances

Asymptotic variances are needed to obtain estimates by the diagonally weighted least-squares method (DWLS). These variances are computed by PRELIS and are stored in a file in binary form (see Jöreskog & Sörbom, 1993a–b). To read it, put the line

```
Asymptotic Variances from File filename
```

in the SIMPLIS input file.

6.4.6 Selection of Variables

Selection of variables is automatic in the sense that only the variables involved in the model are used in the analysis even if there are more variables in the data. One can read data on any number of variables and analyze any subset of these. If K variables are specified as Observed Variables or Labels (see Section 6.3) and only k of these variables are included in the model, the program automatically eliminates the rows and columns of the covariance or correlation matrix corresponding to the $K - k$ remaining variables. This elimination of variables is also done in the asymptotic covariance matrix (see Section 6.4.5). One can also reorder the observed K variables before elimination, see Example 8 in Section 1.5.2. Examples 12A and 12B in Section 2.2 and Example 16A in Section 2.4 illustrate the convenience of automatic selection of variables (see also comments at the end of Section 1.1.2).

6.5 Sample Size

The sample size is the number of cases on which the covariance or correlation matrix is based. Enter the sample size as an integer on the same line or on a new line after the header line with the words Sample Size. You may also include an equals sign or a colon. If a colon is used, there must be a blank space after it but not before it. Thus, the following ways of specifying the sample size are all equivalent.

```
Sample Size
768

Sample Size 768
```

```
Sample Size = 768
```

```
Sample Size: 768
```

The following will also work:

```
Sample Size is 768
```

The sample size is needed to compute standard errors, t-values of parameter estimates, goodness-of-fit measures, and modification indices. *If the sample size is not specified in the input file, the program will stop.*

6.6   Latent Variables or Unobserved Variables

The header for latent variables is `Latent Variables` or `Unobserved Variables`. After the header, enter a list of names of the latent variables. Labels for latent variables are entered in free format in the same way as for observed variables. A name for a latent variable must not be the same as any name of the observed variables.

If there are no latent variables in your model, this part of the input file is omitted.

6.7   Relationships

The header for relationships is `Relationships`, `Relations`, or `Equations`. This header is optional, i.e., relationships can be entered without a header line.

For each relationship, enter a list of variables of the form

   left-hand variable = *right-hand variables*

where *left-hand variable* is a name of a dependent variable and *right-hand variables* is a list of all variables on which the *left-hand variable* depends. The variable names are separated by spaces or + signs.

In a path diagram, a *left-hand variable* is a variable (observed or latent) such that one or more one-way (unidirected) arrows are pointing to it, and the *right-hand variables* are the variables where these arrows are coming from. As a mnemonic, we may also consider the relationship as

   TO *variable* = FROM *variables*

Each relationship begins on a new line. Any number of spaces can be used before and after the equals sign (=). If the right-hand variables are *consecutive variables*, one can use the construction

left-hand variable = VarA - VarK.

to mean that all consecutive variables from VarA to VarK inclusive, are right hand variables. If several consecutive left-hand variables have the same right-hand variables, one can use the construction

Vara - Varm = *right-hand variables*

to specify several relationships simultaneously. This means that each of the consecutive variables from Vara to Varm depend on the same right-hand variables. Any number of spaces may be used before or after the dash (-).

If a relationship does not fit on a single line, continue on the next line by giving the same left-hand variable(s) again and the remaining right-hand variables.

The most general form of a relationship is

varlist1 = *varlist2*

meaning that *each* variable in *varlist1* depends on *all* variables in *varlist2*. Each of the two varlists are lists of variables separated by spaces. Within each varlist, consecutive variables may be specified by means of a dash (-).

The examples of Chapters 1 and 2 illustrate how these rules are applied in practice.

6.8   Paths

The relationships in a model may be specified in terms of paths instead of relationships. In a certain sense, paths are just the opposites of relationships. For example,

FROM *variable* -> TO *variables*

means that there is a path (a one-way arrow) from the FROM *variable* to each of the TO *variables*. The most general form is

FROM *variables* -> TO *variables*

where both FROM *variables* and TO *variables* are lists of variables separated by spaces. This means that there is a path from *each* variable in the FROM *variables* to *all* variables in the TO *variables*. For further clarification of paths, see Example 4 of Chapter 1.

6.9 Scaling the Latent Variables

Latent variables are unobservable and have no definite scales. Both the origin and the unit of measurement in each latent variable are arbitrary. To define the model properly, the origin and the unit of measurement of each latent variable must be defined. For single-sample problems, as in Chapter 1, the origin is usually fixed by assuming that each latent variable has zero mean. In order to interpret all the parameters, units of measurements of the latent variables must also be defined.

Typically, there are two ways in which this is done. The most useful and convenient way of assigning the units of measurement of the latent variables is to assume that they are standardized so that they have unit variances in the population. This means that the unit of measurement in each latent variable equals its population standard deviation. A new feature in LISREL 8 is that this can be done also for the dependent latent variables (Eta-variables). Previously, this was only possible for the independent latent variables (Ksi-variables).

Another way to assign a unit of measurement for a latent variable, is to fix a non-zero coefficient (usually one) in the relationship for one of its observed indicators. This defines the unit for each latent variable in relation to one of the observed variables, a so-called *reference variable*. In practice, one chooses as reference variable the observed variable, which, in some sense, best represents the latent variable. For example, if READING and WRITING are observed indicators of the latent variable Verbal, and we choose READING as the reference variable, then the two measurement relationships are specified as

```
READING = 1*Verbal
WRITING = Verbal
```

This makes READING a reference variable for Verbal. The coefficient of READING on Verbal will not be estimated and will be fixed equal to one. The coefficient of WRITING on Verbal will be estimated, however.

LISREL 8 defines the units of latent variables as follows

☐ If a reference variable is assigned to a latent variable by the specification of a fixed non-zero coefficient in the relationship between the reference variable and the latent variable, then this defines the scale for that latent variable. This holds regardless of whether this latent variable is a dependent latent variable (an η-variable) or an independent latent variable (a ξ-variable).

❏ If no reference variable is specified for a latent variable, by assigning a fixed non-zero coefficient for an observed variable, the program will standardize this latent variable. This holds regardless of whether this latent variable is a dependent latent variable (an η-variable) or an independent latent variable (a ξ-variable).

If a reference variables solution is specified, a standardized solution can always be obtained in LISREL format by putting SS or SC on the LISREL Output line, see Chapter 5.

Further considerations are necessary when the observed variables in the model are ordinal, because then, even the observed variables do not have any units of measurement (see Section 1.6). Other considerations are necessary in longitudinal and multiple group studies in order to put the variables on the same scale at different occasions and in different groups (see Chapter 2). See also Chapter 2 on how to relax the assumption of zero means for the latent variables.

6.10 Starting Values

As mentioned in Section 4.3, the parameters of the LISREL model are estimated by means of an iterative procedure that requires starting values. Since these starting values are normally generated by LISREL 8, for most problems there is no need for users to specify starting values. However, if a good estimate of a parameter is available *a priori*, it is possible to specify this, so that LISREL 8 can make use of it. Such a starting value may be specified for any coefficient in relationships by putting the starting value *in parentheses* followed by an asterisk (*) and then the name of a variable. Thus, in

```
READING = 1*Verbal
WRITING = (1)*Verbal
```

the 1 in the first equation is a fixed coefficient whereas the (1) in the second equation is a starting value.

6.11 Error Variances and Covariances

As mentioned in Section 5.2, there are three kinds of error terms in the full LISREL model:

◻ Measurement errors in observed x-variables.

◻ Measurement errors in observed y-variables.

◻ Structural errors in the structural equations.

It is convenient to refer to these as errors in x, errors in y, and errors in Eta, respectively.

6.11.1 Fixed Error Variances

By default, the variances of all error terms are free parameters to be estimated. Occasionally, it may be necessary to set an error variance to zero or some other fixed value as in Example 7 of Chapter 1. This is done by the command

```
Let the Error Variance of VarA be  a
```

or

```
Set the Error Variance of VarA equal to  a
```

where a is a number.

If VarA - VarH are consecutive variables, one can set all their error variances equal to the same number a by

```
Let the Error Variances of VarA - VarH be  a
```

or

```
Set the Error Variances of VarA - VarH to  a
```

Error variances for two or more variables may also be specified to be equal, see Section 6.13.2.

6.11.2 Error Covariances

All error terms are assumed to be uncorrelated by default, but it is possible to specify correlated errors between two x-variables, between two y-variables, between one x-variable and one y-variable, and between two Eta-variables. It is not permitted to specify correlated errors between an Eta-variable and a x- or y-variable. The general syntax for doing this is

```
Let the Errors between VarA and VarB Correlate
```
or
```
Set the Error Covariance between VarA and VarB Free
```

The parameter estimated will in general be the covariance between the error terms of VarA and VarB. Examples 2, 6, and 9 of Chapter 1 illustrate this.

If there are many error covariances to estimate, one can use

```
Let the Errors between VarA - VarH and VarK - VarZ Correlate
```
or
```
Set the Error Covariances of VarA - VarH and VarK - VarZ Free
```

to set all the error covariances between one set of consecutive variables VarA - VarH and another set of consecutive variables VarK - VarZ free.

One can also set an error covariance to zero (or some other value) with:

```
Set the Error Covariance between VarA and VarB to 0
```

Since the error covariance is zero by default, this is normally not needed, but it may be necessary in multigroup problems if the error covariance between VarA and VarB should be zero in the current group but was free in the previous group (see Example 11 of Chapter 2).

6.12   Uncorrelated Factors

The independent latent variables (the Ksi-variables) are freely correlated by default. If the scales of the latent independent variables are defined by reference variables (see Section 6.9), the variances and covariances of the latent independent variables will be estimated as free parameters. If the latent independent variables are standardized (see Section 6.9), the correlations of the latent independent variables will be estimated as free parameters. Occasionally, one may want to specify that the latent independent variables should be uncorrelated (uncorrelated or orthogonal factors). This is done with the line

```
Set the Covariances of Ksi1 - Ksi7 to 0
```
or
```
Set the Correlations of Ksi1 - Ksi7 to 0
```

6.13 · Equality Constraints

6.13.1 Equal Paths

A path coefficient may be specified to be equal to another path coefficient. This means that the two path coefficient will be treated as a single free parameter rather than as two independent parameters. The SIMPLIS syntax for doing this is:

```
Set the Path from VarA to VarB Equal to the Path from VarC to VarD
```

The following shorter versions of this can also be used.

```
Set Path from VarA to VarB = Path from VarC to VarD
Set Path VarA -> VarB = Path VarC -> VarD
Set VarA -> VarB = VarC -> VarD
```

Using Let lines instead of Set lines, this can also be expressed as:

```
Let the Path from VarA to VarB be Equal to the Path from VarC to VarD
Let Path from VarA to VarB = Path from VarC to VarD
Let Path VarA -> VarB = Path VarC -> VarD
Let VarA -> VarB = VarC -> VarD
```

6.13.2 Equal Error Variances

Two error variances may be specified to be equal with the line

```
Set the Error Variances of VarA and VarB Equal
```

or

```
Let the Error Variances of VarA and VarB be Equal
```

If VarA - VarH are consecutive variables, all their error variances may be set equal with the line:

```
Equal Error Variances: VarA - VarH
```

Note that this is a different type of line than above. The most general form of this is:

```
Equal Error Variances: varlist
```

6.13.3 Freeing a Fixed Parameter or Relaxing an Equality Constraint

The following options are only needed in multigroup problems. Recall that in multigroup problems, all parameters in the model are constrained to be equal across groups unless otherwise specified (see Chapter 2). A parameter that is fixed or constrained may be set free. For paths, the easiest way to do this is to add a relationship (see Example 12 of Chapter 2). For example, if the path from VarA to VarB is fixed or constrained, it can be freed by adding the line

```
VarA -> VarB
```

to the Paths or by adding the line

```
VarB = VarA
```

to the Relationships.

Similarly, if the error variance of VarA is fixed or constrained, it can be freed by adding the line

```
Set the Error Variance of VarA Free
```

or

```
Let the Error Variance of VarA be Free
```

6.14 Options

There are a number of options available to specify the way the output file will be printed and to request additional information in the output file. Each option can either be spelled out directly on a separate line or be put as a two-character keyword on an Options line. For example, to obtain three decimals in the output file, write

```
Number of Decimals = 3
```

or write

```
Options: ND=3
```

The options available are:

```
Print Residuals
Wide Print
Number of Decimals = n
Method of Estimation = Generalized Least Squares
Admissibility Check = Off
Iterations = n
Save Sigma in File filename
```

By putting them on the Options line, they can all be requested on a single line:

```
Options: RS WP ND=3 ME=GLS AD=OFF IT=57 SI= filename
```

These options will now be described. Some of them were illustrated in the Examples in Chapter 1.

6.14.1   Wide Print

By default, the output file will be printed with a maximum of 80 characters per line. The Wide Print or WP option requests that the output file be printed with a maximum of 132 characters per line. If your printer can print wide lines (many printers can be set to print 16 characters per inch), this is useful for reading long equations and large matrices.

To view an output file which has been produced with the WP option, use DISP and the right arrow key to see the text beyond column 80.

6.14.2   Print Residuals

By default, the information contained in residuals and standardized residuals will be given in summary form in the output file (see Example 4 of Chapter 1). The Print Residuals or RS option requests that all residuals be given in matrix form (see Example 5 of Chapter 1 and Section 5.7). The fitted covariance or correlation matrix will also be printed, as well as a Q-plot of the standardized residuals (see Section 5.7).

6.14.3   Number of Decimals

The default number of decimals in the output file is two decimals. This may be changed by writing

```
Number of Decimals = n
```

or by putting ND = n on the Options line, where n is the number of decimals required. Values $1, 2, \ldots, 9$ are permitted.

The term *Number of Decimals* has a special meaning in LISREL 8. If the number of significant digits is at least n, n decimals will be used. However, if the number of significant digits is *less* than n, the number will be printed with as many decimals as is necessary to give n significant digits. For example, if n = 3, numbers such as 0.123, 1.234, and 12.345 will be given with three decimals. But numbers such as 0.0123 and 0.00123 will be given more than three decimals.

6.14.4   Method of Estimation

The parameters of the LISREL model may be estimated by seven different methods:

- Instrumental Variables (IV)
- Two-Stage Least Squares (TSLS)
- Unweighted Least Squares (ULS)
- Generalized Least Squares (GLS)
- Maximum Likelihood (ML)
- Generally Weighted Least Squares (WLS)
- Diagonally Weighted Least Squares (DWLS)

Under general assumptions, all seven methods give consistent estimates of parameters. This means that they will be close to the true parameter values in large samples (assuming, of course, that the model is correct). Technical descriptions of the estimation methods are given in Jöreskog & Sörbom (1989). The seven types of estimates differ in several respects. The TSLS and IV methods are procedures which are non-iterative and very fast. They are used to compute starting values for the other methods, but can also be requested as final estimates. The ULS, GLS, ML, WLS, and DWLS estimates are obtained by means of an iterative procedure that minimizes a particular fit function by successively improving the parameter estimates. WLS requires an estimate of the asymptotic covariance matrix of the sample variances and covariances or correlations being analyzed. Similarly, DWLS requires an estimate of the

asymptotic variances of the sample variances and covariances or correlations being analyzed. These asymptotic variances and covariances are
obtained by PRELIS (Jöreskog & Sörbom, 1993a), which saves them in a
file to be read by LISREL.

If no asymptotic covariance matrix or asymptotic variances is provided, the ML method will be used by default. If an asymptotic covariance
matrix has been read, the WLS method is used by default. If asymptotic
variances have been read, the DWLS method is used by default. To request a method other than the default method, write one of the following
lines

```
Method: Unweighted Least-Squares
Method: Generalized Least-Squares
Method: Two-Stage Least-Squares
Method: Instrumental Variables Method
```

or put the corresponding option on the Options line (only one of the lines
should be included in the input file):

```
Options: ... UL ...
Options: ... GL ...
Options: ... TS ...
Options: ... IV ...
```

The dots (...) indicate that there may be other options or keywords on
the Options line.

6.14.5 Admissibility Check

There is a built-in admissibility check that will stop the iterations after
a specified number of iterations (default = 20) if the solution becomes
non-admissible. The *admissibility* check is that:

□ The matrices LAMBDA-Y and LAMBDA-X have full column ranks
   and no rows of only zeros.

□ All covariance matrices are positive definite.

A failed admissibility check may be an indication of a bad model. A
mistake was made so that the model being estimated differs from the one
intended, or else the model strongly disagrees with the data. The purpose of the admissibility check is to prevent the program from running
for many iterations without producing any useful results. Although there

may be exceptions, our experience suggests that if a solution is not admissible after 20 iterations, it will remain non-admissible if the program is allowed to continue to iterate. The error message that stops the program when the model is not admissible is:

```
F_A_T_A_L  E_R_R_O_R : Admissibility test failed.
```

There will also be warning messages in the output file indicating for which parameter matrix the test fails.

If this message occurs, one should check the data and the model carefully to see if they are as intended. If they are, and the admissibility test still fails after 20 iterations, one can change the value 20 to some larger value such as 30 or 40.

There are situations where one wants to have a non-admissible solution intentionally, so there are provisions to turn the admissibility check off. This occurs, for example, when one specifies an error variance to be zero intentionally.

To set the admissibility check to 30, say, write:

```
Admissibility Check = 30
```

Similarly, to set the admissibility check off, write

```
Admissibility Check = Off
```

To make these specifications on the Options line, write:

```
Options: ... AD=30 ...
```

Similarly, to set the admissibility check off, write

```
Options: ... AD=OFF ...
```

6.14.6 Maximum Number of Iterations

The maximum number of iterations allowed is three times the number of parameters in the model. Our experience is that, for models which are reasonable for the data, the iterations will converge before this maximum is reached. However, for special problems, more iterations may be needed. To set the maximum number of iterations to 100, say, write

```
Iterations = 100
```
or
```
Options: ... IT=100 ...
```

6.14.7 Save Sigma

The fitted covariance matrix, called Sigma, may be saved in a file with the line

`Save Sigma in File` *filename*

or with

`Options: ...` `SI=` *filename* `...`

See the next section for a special use of this option.

6.15 Cross-Validation

The single-sample cross-validation index of Browne & Cudeck (1989) is

$$\text{CVI} = F(\mathbf{S}_v, \hat{\mathbf{\Sigma}}_m) - (1/2n_v)k(k+1)) \tag{6.1}$$

where F is the fit function, $\mathbf{S}_v$ is the covariance matrix or correlation matrix of the validation sample based on sample size $n_v + 1$ and $\hat{\mathbf{\Sigma}}_m$ is the covariance (correlation) matrix fitted in the exploration sample under the model. The last matrix is saved in a file by including the line

`Save Sigma in File` *filename*

in the input file. There will be one Sigma for each model to be cross-validated. For example, to save Sigma for Panel Model 2 (Example 9B) in file **SIGMA2**, include the line

`Save Sigma in File SIGMA2`

in the input file **EX9B.SPL**.

The cross-validation index for Panel Model 2 is then obtained using the following input file:

```
Cross-Validating Panel Model 2
Observed Variables from File PANEL.LAB
Correlation Matrix from File PANELUSA.PMV
Asymptotic Covariance Matrix from file PANELUSA.ACP
Sample Size 395
Crossvalidate File SIGMA2
End of Problem
```

Here, **PANELUSA.PMV** contains $\mathbf{S}_v$, and **SIGMA2** contains $\hat{\boldsymbol{\Sigma}}_m$ fitted in the exploration sample under the model. The sample size 395 is the size of the validation sample on which $\mathbf{S}_v$ is based. The output gives a cross-validation index and a 90 percent confidence interval for this (Browne, 1992).

For the example above the output looks like

```
CROSS-VALIDATION INDEX (CVI) = 0.62
90 PERCENT CONFIDENCE INTERVAL FOR CVI = (0.49 ; 0.77)
```

6.16   LISREL Output

As explained in Chapter 5, one can choose to obtain the estimated LISREL solution either in SIMPLIS output or in LISREL output. In SIMPLIS output, the estimated model is presented in equation form, while in LISREL output the model is presented in matrix form. Otherwise, the two output formats contain the same information. SIMPLIS output will be obtained by default. To request output in LISREL format, put the line

```
LISREL Output
```

in the input file. If nothing else is included on the LISREL Output line, one will obtain the same information as with SIMPLIS output. But it is also possible to obtain additional information by putting various two-character keywords on the LISREL Output line. For example,

```
LISREL Output: SS SC EF SE VA MR FS PC PT
```

Some of these were explained in Chapter 5; others are explained in detail in Jöreskog & Sörbom (1989). A brief description of all keywords follows.

SS Print standardized solution

SC Print completely standardized solution

EF Print total and indirect effects, their standard errors, and t-values

VA Print variances and covariances

MR Equivalent to **RS** and **VA**

FS Print factor scores regression

PC Print correlations of parameter estimates

PT Print technical information

6.17   End of Problem

To indicate the end of the problem, write the line

```
End of Problem
```

This is optional but recommended, especially when several problems are stacked together in the same input file. In multi-sample problems one puts End of Problem after the last group, *not after each group*.

6.18   Additional Command Lines

Since the first release of LISREL 8 with the SIMPLIS syntax, additional lines have been added to accomodate new developments in the program.

Principal Components, Exploratory Factor Analysis, and Regression

The Principal Components line

```
Principal Components with k Components
```

and the command for exploratory factor analysis

```
Factor Analysis with k Factors
```

have been added with version 8.3. Examples of the new Regress line are:

```
Regress Y1 on Y2 and X1 with X1-X3 as Instrumental Variables
Regress Y1 on Y2 X2 and X3 with X1-X3 as Instrumental Variables
```

See also Chapter 3 of *LISREL 8: New Statistical Features*.

Missing Value Code

Full information maximum likelihood (FIML) is automatically invoked when the SIMPLIS command file includes a line

```
Missing Value Code value
```

before the Sample Size and Raw Data lines, where *value* can be any number. See also the *Interactive LISREL: User's Guide*.

System File and PSFFile

```
System File from File ORD31.DSF
```

replaces lines 1–7 in, for example:

```
SIMPLIS File for Estimating the Panel Model
Observed Variables:                                        !   1
NOSAY1 COMPLEX1 NOCARE1 TOUCH1 INTERES1                     !   2
NOSAY2 COMPLEX2 NOCARE2 TOUCH2 INTERES2                     !   3
Means from File PANUSA.ME                                   !   4
Covariance Matrix from File PANUSA.CM                       !   5
Asymptotic Covariance Matrix from File PANUSA.ACC           !   6
Sample Size: 849                                            !   7
Latent Variables: Efficac1 Respons1 Efficac2 Respons2      !   8
...
```

The system file (**DSF** file) has all the information about the variables and the data, even the location of the asymptotic covariance matrix.

The PSFFile command looks like:

```
PSFfile filename.PSF
```

For example,

```
PSFfile KJUDD.PSF
```

See also Chapter 3 of *LISREL 8: New Statistical Features.*

No x-variables and Equal Variances

Simplex and other such models, that is a model with only y- and η-variables (LISREL submodel 3B, see Jöreskog & Sörbom, *LISREL 8: User's Reference Guide*, 1999, Chapter 6) can be specified in SIMPLIS with

```
No x-variables
```

A test of equality of factor variances can now be obtained by adding, for example, the lines (see **ORD33A.SPL**):

```
Equal Variances: Efficac1 Efficac2
Equal Variances: Respons1 Respons2
```

Both additions are described in a document *Structural Equation Modeling with Oridnal Variables using LISREL*[1]

$CLUSTER and $PREDICT

The $CLUSTER and $PREDICT commands were added when multilevel SEM was added to LISREL with version 8.5. See the online Help file for details.

---

[1]Available on SSI's website.

7 COMPUTER EXERCISES

Exercise 1

Kerckhoff (1974, p. 46) reports the correlations in Table 7.1 between a number of variables for 767 twelfth-grade males.

Table 7.1
Correlations for Background, Aspiration, and Attainment Variables

	x_1	x_2	x_3	x_4	y_1	y_2	y_3
INTELLNC	1.000						
SIBLINGS	−.100	1.000					
FATHEDUC	.277	−.152	1.000				
FATHOCCU	.250	−.108	.611	1.000			
GRADES	.572	−.105	.294	.248	1.000		
EDUCEXP	.489	−.213	.446	.410	.597	1.000	
OCCUASP	.335	−.153	.303	.331	.478	.651	1.000

The variables, in the order in which they appear in the table, are:

$x_1 =$ intelligence

$x_2 =$ number of siblings

$x_3 =$ father's education

$x_4 =$ father's occupation

$y_1 =$ grades

$y_2 =$ educational expectation

y_3 = occupational aspiration

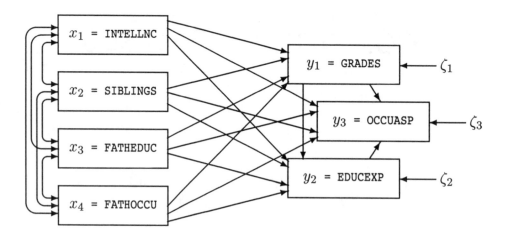

Figure 7.1   Model for Educational and Occupational Aspiration

Problems

Consider a full recursive model in which educational expectation depends on grades and occupational aspiration depends on grades and educational expectation, each dependent variable possibly depending on all the explanatory variables as shown in Figure 7.1. Test each of the following hypotheses:

1. GRADES does not depend on FATHEDUC and FATHOCCU.
2. SIBLINGS does not have any direct effect on any of the dependent variables.
3. There are no direct effects of any explanatory variables on OCCU-ASP. These effects are only mediated via GRADES and EDUCEXP.

Exercise 2

Warren, White, & Fuller (1974) report on a study wherein a random sample of 98 managers of farmer cooperatives operating in Iowa was selected with the objective of studying managerial behavior.

The role behavior of a manager in farmer cooperatives, as measured by his role performance (y), was assumed to be linearly related to the four variables:

x_1 = Knowledge of economic phases of management directed toward profit-making in a business and product knowledge

x_2 = Value orientation: tendency to evaluate rationally means to an economic end

x_3 = Role satisfaction: gratification obtained by the manager from performing the managerial role

x_4 = Past training: amount of formal education

To measure x_1 (knowledge), sets of questions were formulated by specialists in the relevant fields of economics and fertilizers and chemicals. The measure of rational value orientation to economic ends (x_2) was a set of 30 items administered to respondents with a questionnaire in which the respondents were asked to indicate the strength of their agreement or disagreement. The respondents also indicated the strength of satisfaction or dissatisfaction (x_3) for each of 11 statements covering four areas of satisfaction: (1) managerial role itself, (2) the position, (3) rewards, and (4) performance of complementary role players. The amount of past training (x_4) was the total number of years of formal schooling divided by six.

Role performance (y) was measured with a set of 24 questions covering the five functions of planning, organizing, controlling, coordinating, and directing. The recorded verbal responses of managers on how they performed given tasks were scored by judges on a scale of 1 to 99 on the basis of performance leading to successful management. Responses to each question were randomly presented to judges and the raw scores were transformed by obtaining the "Z" value for areas of 0.01 to 0.99 from a cumulative normal distribution (a raw score of 40 received a transformed score of -0.253). For each question, the mean of transformed scores of judges was calculated.

The covariance matrix of the five variables is given in Table 7.2.

Problem A

Estimate the ordinary least squares (OLS) regression of y on x_1, x_2, x_3, x_4. Which of the four variables has a significant effect on y?

Table 7.2   Covariance Matrix for Role Behavior Variables

	y	x_1	x_2	x_3	x_4
ROLBEHAV	0.0209				
KNOWLEDG	0.0177	0.0520			
VALORIEN	0.0245	0.0280	0.1212		
ROLSATIS	0.0046	0.0044	−0.0063	0.0901	
TRAINING	0.0187	0.0192	0.0353	−0.0066	0.0946

Problem B

Suppose the x-variables contain measurement error. Assuming that the reliabilities of the x-variables are

$$0.60, \quad 0.64, \quad 0.81, \quad 1.00,$$

which we take to be known values, re-estimate γ to reduce or eliminate the effects of measurement errors by using a model of the form

$$\mathbf{x} = \boldsymbol{\xi} + \boldsymbol{\delta}$$

$$y = \boldsymbol{\gamma}'\boldsymbol{\xi} + \zeta,$$

where $\boldsymbol{\xi}$ is a vector of true measurements and $\boldsymbol{\delta}$ is a vector of errors of measurement. Which variables do now have significant effects?

Problem C

The analysis in Problem B is problematic because the error variances must be known *a priori*. These error variances cannot be estimated from the information provided by the covariance matrix in Table 7.2 if there is only one measure x for each ξ. However, for each ξ for which there are two or more indicators, one can estimate the measurement errors as well as the structural parameters directly from the data. This is illustrated in the following continuation of the exercise.

To estimate the effect of measurement error in the observed variables, Rock, *et al.* (1977) split each of the measures y, x_1, x_2, and x_3 randomly into two parallel halves. The full covariance matrix of all the split-halves is given in Table 7.3, where the number of items in each split-half is given in parentheses.

Table 7.3 Covariance Matrix for Split-halves Variables

		y_1	y_2	x_{11}	x_{12}	x_{21}	x_{22}	x_{31}	x_{32}	x_4
y_1	(12)	.0271								
y_2	(12)	.0172	.0222							
x_{11}	(13)	.0219	.0193	.0876						
x_{12}	(13)	.0164	.0130	.0317	.0568					
x_{21}	(15)	.0284	.0294	.0383	.0151	.1826				
x_{22}	(15)	.0217	.0185	.0356	.0230	.0774	.1473			
x_{31}	(5)	.0083	.0011	−.0001	.0055	−.0087	−.0069	.1137		
x_{32}	(6)	.0074	.0015	.0035	.0089	−.0007	−.0088	.0722	.1024	
x_4		.0180	.0194	.0203	.0182	.0563	.0142	−.0056	−.0077	.0946

This can be used to estimate the true regression equation

$$\eta = \gamma_1 \xi_1 + \gamma_2 \xi_2 + \gamma_3 \xi_3 + \gamma_4 \xi_4 + \zeta \tag{7.1}$$

using the following measurement models.

$$\begin{pmatrix} y_1 \\ y_2 \end{pmatrix} = \begin{pmatrix} 1 \\ 1 \end{pmatrix} \eta + \begin{pmatrix} \epsilon_1 \\ \epsilon_2 \end{pmatrix} \tag{7.2}$$

$$\begin{pmatrix} x_{11} \\ x_{12} \\ x_{21} \\ x_{22} \\ x_{31} \\ x_{32} \\ x_4 \end{pmatrix} = \begin{pmatrix} 1 & 0 & 0 & 0 \\ 1 & 0 & 0 & 0 \\ 0 & 1 & 0 & 0 \\ 0 & 1 & 0 & 0 \\ 0 & 0 & 1 & 0 \\ 0 & 0 & 1.2 & 0 \\ 0 & 0 & 0 & 1 \end{pmatrix} \begin{pmatrix} \xi_1 \\ \xi_2 \\ \xi_3 \\ \xi_4 \end{pmatrix} + \begin{pmatrix} \delta_{11} \\ \delta_{12} \\ \delta_{21} \\ \delta_{22} \\ \delta_{31} \\ \delta_{32} \\ 0 \end{pmatrix} \tag{7.3}$$

The value 1.2 in the last equation reflects the fact that x_{32} has six items whereas x_{31} has only five.

The latent variables are:

$\eta =$ role behavior

$\xi_1 =$ knowledge

$\xi_2 =$ value orientation

$\xi_3 =$ role satisfaction

$\xi_4 =$ past training

The observed variables are:

$y_1 =$ a split-half measure of role behavior

$y_2 =$ a split-half measure of role behavior

$x_{11} =$ a split-half measure of knowledge

$x_{12} =$ a split-half measure of knowledge

$x_{21} =$ a split-half measure of value orientation

$x_{22} =$ a split-half measure of value orientation

$x_{31} =$ a split-half measure of role satisfaction

$x_{32} =$ a split-half measure of role satisfaction

$x_4 = \xi_4 =$ a measure of past training

Estimate the regression equation (7.1). Which variables do now have significant effects? Does the model (7.1) – (7.3) fit the data? If so, test the hypothesis $\mathsf{Var}(\epsilon_1) = \mathsf{Var}(\epsilon_2)$ and $\mathsf{Var}(\delta_{i1}) = \mathsf{Var}(\delta_{i2})$, $i = 1, 2$. Using these results, estimate the reliabilities of the original variables y, x_1, x_2, and x_3 in Table 7.2 and compare them with the ones assumed known in Problem B.

Exercise 3

Wiley & Hornik (1973) give data from a study of communication processes conducted in El Salvador. The two construct variables are television watching by children (η) and television possession by their family (ξ). Each construct was measured by two congeneric measures at three points in time. The covariance matrix ($N = 189$) of the resulting twelve variables is shown in Table 7.4.

Problems

A Estimate and test the model in Figure 7.2. If the model does not fit the data, determine the major source of lack of fit and modify the model accordingly.

B Examine the evidence for specific factors in each of the four measures.

C Is the measurement model for the two constructs invariant over time?

D Is the structural model for η and ξ invariant over time?

Table 7.4
Covariance Matrix for Two Versions of Television Possession and Watching at Three Occasions

	x_{11}	x_{21}	y_{11}	y_{21}	x_{12}	x_{22}	y_{12}	y_{22}	x_{13}	x_{23}	y_{13}	y_{23}
x_{11}	.247											
x_{21}	.227	.248										
y_{11}	.479	.443	2.313									
y_{21}	.312	.288	1.225	2.264								
x_{12}	.198	.196	.432	.283	.250							
x_{22}	.201	.205	.437	.299	.232	.250						
y_{12}	.402	.375	1.459	1.053	.468	.467	2.453					
y_{22}	.212	.185	.906	1.212	.267	.271	1.028	1.749				
x_{13}	.185	.176	.375	.250	.203	.202	.397	.221	.250			
x_{23}	.176	.172	.351	.247	.196	.199	.381	.210	.233	.250		
y_{13}	.376	.330	1.308	1.020	.418	.419	1.517	.906	.502	.491	2.476	
y_{23}	.231	.204	.907	1.186	.274	.273	1.004	1.107	.319	.312	1.326	2.281

x_{it} = measure i of possession at time t
y_{it} = measure i of watching at time t

Exercise 4

Wehrle (1982) gives data on several measures supposed to measure two theoretical constructs, *Control of Pace* and *Powerlessness*, in a sample of

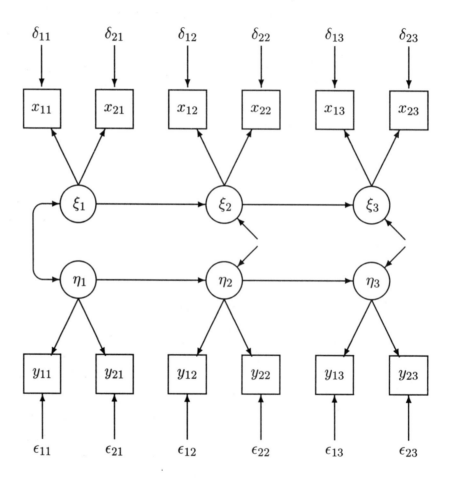

Figure 7.2
Longitudinal Model for Two Versions of Television Possession and Watching at Three Occasions

119 white female secretaries working in the United States. Both theoretical constructs were measured with three different methods so that multiple indicators could be used to test the measurement hypothesis. Furthermore, the study employed a longitudinal design with a lag time of three months. This lag was included so that certain causal models could be tested, but this feature of the design also means that the same latent variables were measured at two points in time — if the hypothesized measurement model holds.

The three measuring methods for the two constructs were: (L) response to a Likert scale composed of five statements expressing progressively greater control over the pace of work or progressively lesser powerlessness at work, (N) a number method where the respondent rated herself on a scale from 1 to 10, and (R) a rating from 1 to 5 made by a trained rater based on the respondent's narrative description of her work.

The variables are

x_{1L} = Control of pace of work at time 1 measured by Likert method

x_{1N} = Control of pace of work at time 1 measured by Number method

x_{1R} = Control of pace of work at time 1 measured by Rating method

x_{2L} = Control of pace of work at time 2 measured by Likert method

x_{2N} = Control of pace of work at time 2 measured by Number method

x_{2R} = Control of pace of work at time 2 measured by Rating method

y_{1L} = Powerlessness at work at time 1 measured by Likert method

y_{1N} = Powerlessness at work at time 1 measured by Number method

y_{1R} = Powerlessness at work at time 1 measured by Rating method

y_{2L} = Powerlessness at work at time 2 measured by Likert method

y_{2N} = Powerlessness at work at time 2 measured by Number method

y_{2R} = Powerlessness at work at time 2 measured by Rating method

and the covariance matrix is given in Table 7.5.

Problems

Analyze the data to answer the questions

1. Do the six variables measure the two constructs they are supposed to measure? If so, how highly correlated are these constructs? Is the measurement of these constructs stable over time?

Table 7.5
Covariance Matrix of Control of Pace and Powerlessness Measured by Three Methods at Two Points in Time

x_{1L}	x_{1N}	x_{1R}	x_{2L}	x_{2N}	x_{2R}	y_{1L}	y_{1N}	y_{1R}	y_{2L}	y_{2N}	y_{2R}
.976											
1.056	3.480										
.339	.699	.517									
.475	.793	.213	1.025								
.949	1.660	.508	1.694	4.625							
.248	.494	.212	.438	.918	.537						
.272	.277	.202	.180	.194	.125	.903					
.756	1.797	.879	.542	1.075	.363	1.522	7.126				
.303	.440	.239	.264	.520	.223	.508	1.577	.933			
.270	.432	.234	.388	.785	.311	.387	1.081	.527	1.117		
.586	1.131	.604	.849	2.087	.794	1.036	3.499	1.363	1.715	6.065	
.340	.603	.238	.315	.690	.341	.441	1.404	.674	.649	1.742	.974

2. Is there a specific method variance in the observed measures attributable to each method? If so, what is the relative variance contribution to the total variance due to trait, method, and error and are these contributions invariant over time?

Exercise 5

McDonald (1985) gives the correlation matrix in Table 7.6 for three measures of verbal ability, V1, V2, V3, three measures of word fluency, W1, W2, W3, and three measures of reasoning ability, R1, R2, R3. The original data are from a study by Thurstone. The sample size is 213.

Problem A

Estimate and test the model in Figure 7.3

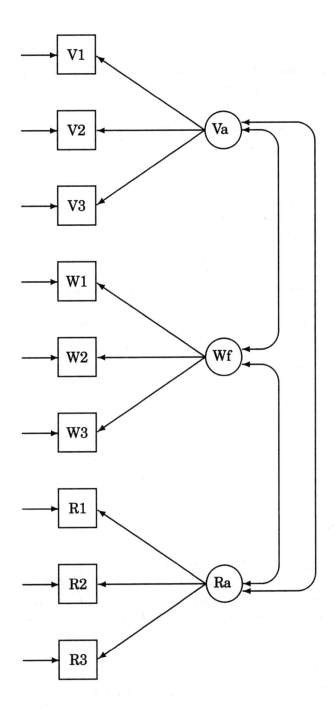

Figure 7.3
Confirmatory Factor Analysis Model for Nine Psychological Tests

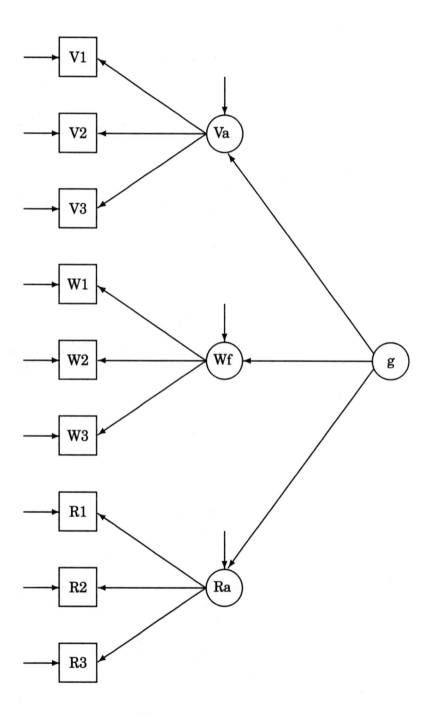

Figure 7.4
Second-Order Factor Analysis Model for Nine Psychological Tests

Table 7.6   Correlation Matrix for Nine Psychological Tests

V1	1.000								
V2	0.828	1.000							
V3	0.776	0.779	1.000						
W1	0.439	0.493	0.460	1.000					
W2	0.432	0.464	0.425	0.674	1.000				
W3	0.447	0.489	0.443	0.590	0.541	1.000			
R1	0.447	0.432	0.401	0.381	0.402	0.288	1.000		
R2	0.541	0.537	0.534	0.350	0.367	0.320	0.555	1.000	
R3	0.380	0.358	0.359	0.424	0.446	0.325	0.598	0.452	1.000

Problem B

Consider the second-order factor analysis model in Figure 7.4. Estimate and test this model in completely standardized form. Explain why the two models have the same fit.

Exercise 6

Hodge & Treiman (1968) studied the relationship between social status and social participation. For a sample of 530 women, they report data on x_1 = income, x_2 = occupation, x_3 = education, y_1 = church attendance, y_2 = memberships, and y_3 = friends seen. All variables are expressed in standardized form. The y's are viewed as independent indicators of a latent variable η = social participation which is caused by the x's. Thus,

$$\eta = \gamma_1 x_1 + \gamma_2 x_2 + \gamma_3 x_3 + \zeta ,$$

$$y_1 = \lambda_1 \eta + \epsilon_1, \quad y_2 = \lambda_2 \eta + \epsilon_2, \quad y_3 = \lambda_3 \eta + \epsilon_3 .$$

From a substantive viewpoint, it may be helpful to view the x's as determining

$$\xi = \gamma_1 x_1 + \gamma_2 x_2 + \gamma_3 x_3 = social\ status ,$$

which in turn determines

$$\eta = \xi + \zeta = social \ participation \ .$$

A path diagram is given in Figure 7.5. The correlations of the variables are given in Table 7.7.

Table 7.7   Correlations for Variables in MIMIC Model

	x_1	x_2	x_3	y_1	y_2	y_3
INCOME	1.000					
OCCUPAT	.304	1.000				
EDUCAT	.305	.344	1.000			
CHURCHAT	.100	.156	.158	1.000		
MEMBERSH	.284	.192	.324	.360	1.000	
FRIENDS	.176	.136	.226	.210	.265	1.000

Problem

Estimate and test the MIMIC model in Figure 7.5. Test the hypothesis that the variance of ζ is zero.

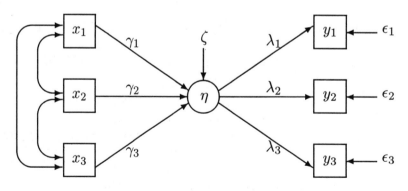

Figure 7.5   Path Diagram for MIMIC Model

Exercise 7

Bollen (1989a, p. 362) gives the covariance matrices shown in Table 7.8 for five variables on perceived and subjective socioeconomic status for 368 blacks and 432 whites. The data come from a study of Kluegel, *et al.* (1977). The variables are:

y_1 = subjective income

y_2 = subjective occupational prestige

y_3 = subjective overall status

x_1 = income

x_2 = occupational prestige

Table 7.8
Covariance Matrices for Five Variables on Perceived and Subjective Socioeconomic Status for Blacks and Whites

Blacks: $N = 368$

	y_1	y_2	y_3	x_1	x_2
y_1	.663				
y_2	.334	.558			
y_3	.301	.282	.615		
x_1	.724	.420	.391	4.397	
x_2	2.417	3.357	1.806	7.485	263.218

Whites: $N = 432$

	y_1	y_2	y_3	x_1	x_2
y_1	.449				
y_2	.166	.410			
y_3	.226	.173	2.393		
x_1	.564	.259	.382	4.381	
x_2	2.366	3.840	3.082	13.656	452.770

Consider the model in Figure 7.6. Here

η_1 = unobserved subjective income

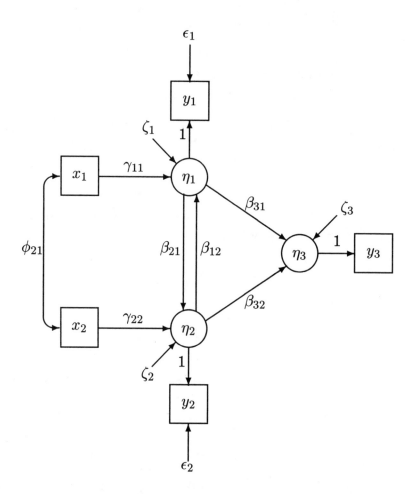

Figure 7.6   Model for Objective and Subjective Socioeconomic Status

$\eta_2 =$ unobserved subjective occupational prestige

$\eta_3 =$ unobserved subjective overall status

Note that there is only one observed indicator of η_1, η_2, and η_3. Nevertheless, verify that one can estimate the measurement error variances of y_1 and y_2 (but not of y_3) as well as the variances of the structural errors ζ_1, ζ_2, and ζ_3.

Problem A

Estimate and test the model for blacks

Problem B

Estimate and test the model for whites

Problem C

Test the hypothesis that the structural parameters γ_{11}, γ_{22}, β_{12}, β_{21}, β_{31}, and β_{32} are the same for blacks and whites.

Exercise 8

Kluegel, Singleton, & Starnes (1977) analyzed data on subjective and objective social class. The data are given in Table 7.9.

There are three variables measuring objective status and four measuring subjective status. The objective status measures are:

□ education — indicated by five categories ranging from less than ninth grade to college graduate.

□ occupation — indicated by the two-digit Duncan SEI score.

□ income — indicated by the total yearly family income before taxes in 1967, coded in units of $2,000 and ranging from under $2,000 to $16,000 or more.

All subjective class indicators were structured questions asking respondents to place themselves in one of four class categories: lower, working, middle, or upper. The questions asked the respondents to indicate which social class they felt their occupation, income, way of life, and influence were most like. The criteria, in terms of which class self-placements were made, correspond directly to the Weberian dimensions of economic class (occupation and income), status (lifestyle), and power (influence).

Table 7.9
Correlations, Means and Standard Deviations for Indicators of Objective Class and Subjective Class.

Correlations for whites below diagonal and for blacks above diagonal

N (Whites) = 432; N (Blacks) = 368

Variables	y_1	y_2	y_3	x_1	x_2	x_3	x_4	Mean (s.d.)
EDUC	—	.404	.268	.216	.233	.211	.207	1.274 (1.106)
OCC	.495	—	.220	.277	.183	.270	.157	2.347 (1.622)
INC	.398	.292	—	.268	.424	.325	.282	4.041 (2.097)
SC-OCC	.218	.282	.184	—	.550	.574	.482	1.288 (0.747)
SC-INC	.299	.166	.383	.386	—	.647	.517	1.129 (0.814)
SC-LST	.272	.161	.321	.396	.553	—	.647	1.235 (0.786)
SC-INF	.269	.169	.191	.382	.456	.534	—	1.318 (0.859)
Mean (s.d.)	1.655 (1.203)	3.670 (2.128)	5.040 (2.198)	1.543 (0.640)	1.548 (0.670)	1.542 (0.623)	1.601 (0.624)	

Problem A

Test the hypothesis that blacks and whites have the same covariance matrix for the observed variables

❑ of the objective measures

□ of the subjective measures

□ of the objective and subjective measures jointly

If any of these hypotheses is rejected, test the less restrictive hypothesis that blacks and whites differ in standard deviations but not in correlations. State the assumptions on which the analysis is based and evaluate the validity of these.

Problem B

Without assuming equality of covariance matrices, test the hypothesis that blacks and whites have the same mean vector for the observed variables

□ of the objective measures

□ of the subjective measures

□ of the objective and subjective measures jointly

Will the conclusions be different if one assumes equality of covariance matrices?

Problem C

Consider Model A in Figure 7.7. This is a measurement model for objective class. Test each of the following hypotheses:

□ Blacks and whites have equal factor loadings.

□ Blacks and whites have equal factor loadings and equal error variances.

□ Blacks and whites have equal factor loadings, equal error variances and equal variance of subjective class.

Adopting one of these hypotheses, estimate the mean difference in subjective class assuming equal or unequal intercepts in the measurement relations. Assuming unequal intercepts, test the hypothesis of no mean difference in objective class.

Problem D

Consider Model B in Figure 7.8. This is a measurement model for subjective class. Test each of the following hypotheses:

- ❑ Blacks and whites have equal factor loadings.
- ❑ Blacks and whites have equal factor loadings and equal error variances.
- ❑ Blacks and whites have equal factor loadings, equal error variances and equal variance of objective class.

Adopting one of these hypotheses, estimate the mean difference in objective class assuming equal or unequal intercepts in the measurement relations. Assuming unequal intercepts, test the hypothesis of no mean difference in subjective class.

Problem E

Consider Model C in Figure 7.9. This is a structural equation model for objective and subjective class.

- ❑ Estimate and test Model C (with both latent variables standardized) for blacks and whites separately.
- ❑ Test the hypothesis of equal factor loadings for blacks and whites for both objective and subjective measures. If this model does not fit the data, verify that the fit can be improved by allowing correlated measurement errors between INCOME and SC-INCM and between OCCUPAT and SC-OCCU.
- ❑ Adopting the previous model, test the hypothesis of equal γ for blacks and whites.
- ❑ Adopting the previous model with equal γ's for blacks and whites, estimate the structural equation

$$subjective\ class = \alpha + \gamma\ objective\ class$$

where α is the mean difference in subjective class after controlling for the difference in objective class. Test the hypothesis that α is zero.

- ❑ Interpret and discuss the substantive meaning of all results.

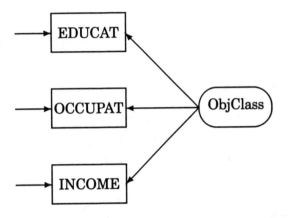

Figure 7.7   Model A: Measurement Model for Objective Class

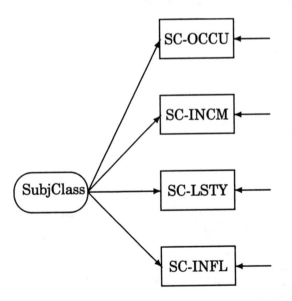

Figure 7.8   Model B: Measurement Model for Subjective Class

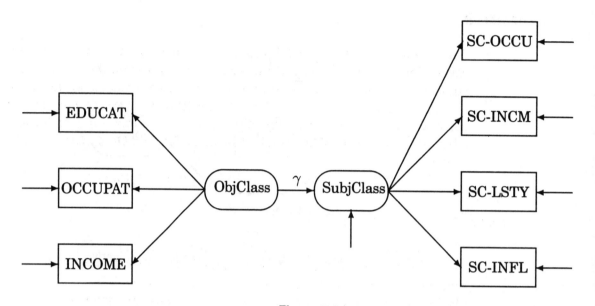

Figure 7.9
Model C: Structural Equation Model for Objective and Subjective Class

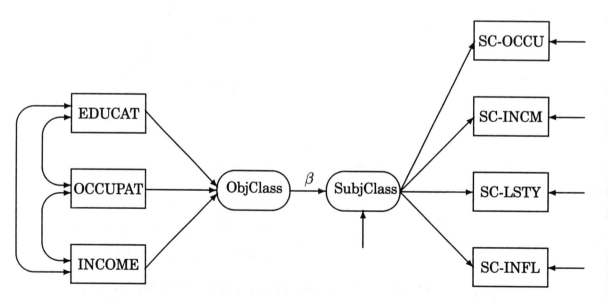

Figure 7.10   Model D: Mimic Model for Objective and Subjective Class

Exercise 9

Some sociologists argue that Education, Occupation, and Income are not caused by Objective Class, but rather the other way around, see, e.g., discussion on cause (formative or induced) and effect (reflective) indicators in Bollen (1989a, pp. 64–67). Regarding Education, Occupation, and Income as cause indicators of Objective Class is the same thing as defining Objective Class as a linear combination of Education, Occupation, and Income:

$$\text{ObjClass} = \alpha_1 \text{EDUCAT} + \alpha_2 \text{OCCUPAT} + \alpha_3 \text{INCOME} . \qquad (7.4)$$

Problems

Using the same data as in the previous exercise, consider Model D in Figure 7.10. Assuming equal factor loadings in the measurement model for Subjective Class, test the hypotheses that the α's in (7.4) are the same for blacks and whites and that β is the same for blacks and whites. Interpret the results and discuss relative advantages of Model C versus Model D.

References

Akaike, H. (1974) A new look at statistical model identification. *IEEE transactions on Automatic Control*, 19, 716–723.

Akaike, H. (1987) Factor analysis and AIC. *Psychometrika*, 52, 317–332.

Aish, A.M., & Jöreskog, K.G. (1990) A panel model for political efficacy and responsiveness: An application of LISREL 7 with weighted least squares. *Quality and Quantity, 24*, 405–426.

Alwin, D.F., & Jackson, D.J. (1981) Applications of simultaneous factor analysis to issues of factorial invariance. Pp. 249–279 in D. Jackson & E. Borgatta (Eds.): *Factor analysis and measurement in sociological research: A Multi-Dimensional Perspective*. Beverly Hills: Sage.

Bagozzi, R.P. (1980c) Performance and satisfaction in an industrial sales force: An examination of their antecedents and simultaneity. *Journal of Marketing*, 44, 65–77.

Bentler, P.M. (1990) Comparative fit indexes in structural models. *Psychological Bulletin*, 107, 238–246.

Bentler, P.M., & Bonett, D.G. (1980) Significance tests and goodness of fit in the analysis of covariance structures. *Psychological Bulletin*, 88, 588–606.

Bentler, P.M., & Woodward, J.A. (1978) A Head Start reevaluation: Positive effects are not yet demonstrable. *Evaluation Quarterly*, 2, 493–510.

Blalock, H.M., Jr., (Ed.) (1985) *Causal models in the social sciences*. Second Edition. New York: Aldine Publishing Co.

Bohrnstedt, G.W. (1969) Observations on the measurement of change. In E.F. Borgatta (Ed.): *Sociological Methodology 1969*. San Francisco: Jossey-Bass, 113–133.

Bollen, K.A. (1986) Sample size and Bentler & Bonett's nonnormed fit index. *Psychometrika*, 51, 375–377.

Bollen, K.A. (1987) Total, direct, and indirect effects in structural equation models. In C. Clogg (Ed.): *Sociological Methodology 1987*. San Francisco: Jossey Bass.

Bollen, K.A. (1989a) *Structural equations with latent variables*. New York: Wiley.

Bollen, K.A. (1989b) A new incremental fit index for general structural equation models. *Sociological Methods and Research*, 17, 303–316.

Bollen, K.A., & Liang, J. (1988) Some properties of Hoelter's CN. *Sociological Methods and Research*, 16, 492–503.

Bollen, K.A., & Long, J.S. (Editors) (1993) *Testing structural equation models*, Sage Publications.

Bozdogan, H. (1987) Model selection and Akaike's information criteria (AIC). *Psychometrika*, 52, 345–370.

Browne, M.W. (1984) Asymptotically distribution-free methods for the analysis of covariance structures. *British Journal of Mathematical and Statistical Psychology*, 37, 62–83.

Browne, M.W. (1992) Confidence interval for CVI. Personal communication.

Browne, M.W., & Cudeck, R. (1989) Single sample cross-validation indices for covariance structures. *Multivariate Behavioral Research*, 24, 445–455.

Browne, M. W. & Cudeck, R. (1993) Alternative ways of assessing model fit. In K. A. Bollen & J. S. Long (Editors): *Testing Structural Equation Models*, Sage Publications.

Calsyn, J.R., & Kenny, D.A. (1977) Self-concept of ability and perceived evaluation of others: Cause or effect of academic achievement? *Journal of Educational Psychology*, 69, 136–145.

Costner, H.L. (1969) Theory, deduction, and rules of correspondence. *American Journal of Sociology*, 75, 245–263.

Cudeck, R. (1989) Analysis of correlation matrices using covariance structure models. *Psychological Bulletin*, 105, 317–327.

Cudeck, R., & Browne, M.W. (1983) Cross-validation of covariance structures. *Multivariate Behavioral Research*, 18, 147–157.

Draper, N.S., & Smith, H. (1967) *Applied regression analysis.* New York: Wiley

Duncan, O.D. (1966) Path analysis: Sociological examples. *American Journal of Sociology,* 72, 1–16.

Duncan, O.D. (1969) Some linear models for two-wave, two-variable panel analysis. *Psychological Bulletin,* 72, 177–182.

Duncan, O.D. (1972) Unmeasured variables in linear models for panel analysis. In H.L. Costner (Ed.): *Sociological Methodology 1972.* San Francisco: Jossey-Bass, 36–82.

Duncan, O.D. (1975) *Introduction to structural equation models.* New York: Academic Press.

Duncan, O.D., Haller, A.O., & Portes, A. (1968) Peer influence on aspiration: A reinterpretation. *American Journal of Sociology,* 74, 119–137.

Finn, J.D. (1974) *A general model for multivariate analysis.* New York: Holt, Reinhart and Winston.

French, J.V. (1951) The description of aptitude and achievement tests in terms of rotated factors. *Psychometric Monographs,* 5.

Goldberger, A.S. (1964) *Econometric theory.* New York: Wiley.

Guilford, J.P. (1956) The structure of intellect. *Psychological Bulletin,* 53, 267–293.

Hodge, R.W., and Treiman, D.J. (1968) Social participation and social status. *American Sociological Review,* 33, 723–740.

Hoelter, J.W. (1983) The analysis of covariance structures: Goodness-of-fit indices. *Sociological Methods and Research,* 11, 325–344.

Huitema, B.H. (1980) *The analysis of covariance and alternatives.* New York: Wiley.

Hauser, R.M., & Goldberger, A.S. (1971) The treatment of unobservable variables in path analysis. Pp. 81–117 in Costner, H.L.: *Sociological Methodology.*

Heise, D.R. (1969) Separating reliability and stability in test-retest correlation. *American Sociological Review,* 34, 93–101.

Heise, D.R. (1970) Causal inference from panel data. In E.F. Borgatta & G.W. Bohrnstedt (Eds.): *Sociological Methodology 1970.* San Francisco: Jossey-Bass, 3–27.

Holzinger, K., & Swineford, F. (1939) *A study in factor analysis: The stability of a bifactor solution*. Supplementary Educational Monograph no. 48. Chicago: University of Chicago Press.

Jagodzinski, W., & Kühnel, S.M. (1988) Estimation of reliability and stability in single-indicator multiple-wave models. *Sociological Methods and Research*, 15, 219–258.

James, L.R., Mulaik, S.A., & Brett, J.M. (1982) *Causal analysis: Assumptions, models, and data*. Beverly Hills: Sage.

Johnston, J. (1972) *Econometric methods*. New York: McGraw-Hill.

Jöreskog, K.G. (1971) Simultaneous factor analysis in several populations. *Psychometrika*, 36, 409–426.

Jöreskog, K.G. (1979a) Basic ideas of factor and component analysis. In K.G. Jöreskog & D. Sörbom: *Advances in factor analysis and structural equation models*. Cambridge, Mass.: Abt Books, 5–20.

Jöreskog, K.G. (1979b) Statistical estimation of structural models in longitudinal developmental investigations. In J.R. Nesselroade & P.B. Baltes (Eds.): *Longitudinal research in the study of behavior and development*. New York: Academic Press.

Jöreskog, K.G. (1990) New developments in LISREL: Analysis of ordinal variables using polychoric correlations and weighted least squares. *Quality and Quantity*, 24, 387–404.

Jöreskog, K.G. (1993) Testing structural equation models. In K.A. Bollen & J.S. Long (Eds), *Testing Structural Equation Models*. Sage Publications.

Jöreskog, K.G., & Sörbom, D. (1976) Statistical models and methods for test-retest situations. In D.N.M. deGruijter & L.J.Th. van der Kamp (Eds.): *Advances in psychological and educational measurement*. New York: Wiley, 285–325.

Jöreskog, K.G., & Sörbom, D. (1977) Statistical models and methods for analysis of longitudinal data. In D.J. Aigner & A.S. Goldberger (Eds.): *Latent variables in socioeconomic models*. Amsterdam: North-Holland Publishing Co., 285–325.

Jöreskog, K.G., & Sörbom, D. (1982) Recent developments in structural equation modeling. *Journal of Marketing Research*, 19, 404–416.

Jöreskog, K.G., & Sörbom, D. (1985) Simultaneous analysis of longitudinal data from several cohorts. Pp. 323–341 in W.M. Mason & S.E. Fienberg (Eds.): *Cohort analysis in social research: Beyond the identification problem*. New York: Springer-Verlag.

Jöreskog, K.G., & Sörbom, D. (1988) *PRELIS - A program for multivariate data screening and data summarization. A preprocessor for LISREL*. Second Edition. Chicago, Illinois: Scientific Software, Inc.

Jöreskog, K.G., & Sörbom, D. (1989a) *LISREL 7 - A guide to the program and applications*. Second Edition. Chicago: SPSS Publications.

Jöreskog, K.G., & Sörbom, D. (1989b) *LISREL 7 User's Reference Guide*. Chicago: Scientific Software International.

Jöreskog, K.G., & Sörbom, D. (1990) Model search with TETRAD II and LISREL. *Sociological Methods and Research*, 19, 93–106.

Jöreskog, K.G., & Sörbom, D. (1993a) New features in PRELIS 2. Chicago: Scientific Software International.

Jöreskog, K.G., & Sörbom, D. (1993) New features in LISREL 8. Chicago: Scientific Software International.

Kaplan, D. (1989) Model modification in covariance structure analysis: Application of the expected parameter change statistic. *Multivariate Behavioral Research*, 24, 285–305.

Kenny, D.A., & Judd, C.M. (1984) Estimating the nonlinear and interactive effects of latent variables. *Psychological Bulletin*, 96, 201–210.

Kerckhoff, A.C. (1974) *Ambition and attainment*. Rose Monograph Series.

Kluegel, J.R., Singleton, R., & Starnes, C.E. (1977) Subjective class identification: A multiple indicators approach. *American Sociological Review*, 42, 599–611.

Lee, S. & Hershberger, S. (1990) A simple rule for generating equivalent models in covariance structure modeling. *Multivariate Behavioral Research*, 25, 313–334.

Lomax, R.G. (1983) A guide to multiple-sample structural equation modeling. *Behavior Research Methods and Instrumentation*, 15, 580–584.

Magidson, J. (1977) Toward a causal model approach for adjusting for pre-existing differences in the non-equivalent control group situation. *Evaluation Quarterly*, 1, 399–420.

Maiti, S.S., & Mukherjee, B.N. (1990) A note on distributional properties of the Jöreskog-Sörbom fit indices. *Psychometrika*, 55, 721–726.

Mare, R.D., & Mason, W.M. (1981) Children's report of parental socioeconomic status. In G.W. Bohrnstedt & E.F. Borgatta (Eds.): *Social Measurement: Current Issues*. Beverly Hills: Sage Publications.

McDonald, R.P. (1985) *Factor analysis and related methods*. Hillsdale, N.J.: Lawrence Erlbaum.

McDonald, R.P. (1989) An index of goodness of fit based on non-centrality. *Journal of Classification*, 6, 97–103.

McDonald, J.A., & Clelland, D.A. (1984) Textile workers and union sentiment. *Social Forces*, 63, 502–521.

McGaw, B., & Jöreskog, K.G. (1971) Factorial invariance of ability measures in groups differing in intelligence and socio-economic status. *British Journal of Mathematical and Statistical Psychology*, 24, 154–168.

Matsueda, R.L., & Bielby, W.T. (1986) Statistical power in covariance structure models. Pp. 120–158 in N.B. Tuma (Ed.): *Sociological Methodology 1986*. San Francisco: Jossey Bass.

Meredith, W. (1964) Rotation to achieve factorial invariance. *Psychometrika*, 29, 187–206.

Mulaik, S., James, L., Van Alstine, J., Bennett, N., Lind, S., & Stilwell, C. (1989) Evaluation of goodness-of-fit indices for structural equation models. *Psychological Bulletin*, 105, 430–445.

Muthén, B. (1984) A general structural equation model with dichotomous, ordered categorical, and continuous latent variable indicators. *Psychometrika*, 49, 115–132.

Olsson, U. (1979) Maximum likelihood estimation of the polychoric correlation coefficient. *Psychometrika*, 44, 443–460.

Rock, D.A., Werts, C.E., Linn, R.L., & Jöreskog, K.G. (1977) A maximum likelihood solution to the errors in variables and errors in equation models. *Journal of Multivariate Behavioral Research*, 12, 187–197.

Saris, W.E., Satorra, A. & Sörbom, D. (1987) The detection and correction of specification errors in structural equation models. In C. Clogg (Ed.): *Sociological Methodology 1987*. San Francisco: Jossey. Bass.

Satorra, A., & Saris, W.E. (1985) Power of the likelihood ratio test in covariance structure analysis. *Psychometrika*, 50, 83–90.

Sobel, M. (1982) Asymptotic confidence intervals for indirect effects in structural equation models. In S. Leinhardt (Ed.): *Sociological Methodology 1982*. San Francisco: Jossey Bass.

Steiger, J.H. (1990) Structural model evaluation and modification: An interval estimation approach. *Multivariate Behavioral Research*, 25, 173–180.

Stelzl, I. (1986) Changing causal relationships without changing the fit: Some rules for generating equivalent LISREL models. *Multivariate Behavioral Research*, 21, 309–331.

Sörbom, D. (1974) A general method for studying differences in factor means and factor structures between groups. *British Journal of Mathematical and Statistical Psychology*, 27, 229–239.

Sörbom, D. (1975) Detection of correlated errors in longitudinal data. *British Journal of Mathematical and Statistical Psychology*, 28, 138–151.

Sörbom, D. (1976) A statistical model for the measurement of change in true scores. In D.N.M. de Gruijter & J.L.Th. van der Kamp (Eds.): *Advances in psychological and educational measurement*. New York: Wiley. 159–169.

Sörbom, D. (1978) An alternative to the methodology for analysis of covariance. *Psychometrika*, 43, 381–396.

Sörbom, D. (1981) Structural equation models with structured means. In K.G. Jöreskog & H. Wold (Eds.): *Systems under indirect observation: Causality, structure and prediction*, Vol. 1. Amsterdam: North-Holland Publishing Co.

Sörbom, D. (1989) Model modification. *Psychometrika*, 54, 371–384.

Sörbom, D., & Jöreskog, K.G. (1981) The use of LISREL in sociological model building. In E. Borgatta & D.J. Jackson (Eds.) *Factor analysis and measurement in sociological research: A multidimensional perspective*. Beverly Hills: Sage.

Tanaka, J.S., & Huba, G.J. (1985) A fit index for covariance structure models under arbitrary GLS estimation. *British Journal of Mathematical and Statistical Psychology*, 42, 233–239.

Theil, H. (1971) *Principles of econometrics*. New York: Wiley.

Thurstone, L.L. (1938) Primary mental abilities. *Psychometric Monographs*, 1.

Tucker, L.R., & Lewis, C. (1973) A reliability coefficient for maximum likelihood factor analysis. *Psychometrika*, 38, 1–10.

Turner, M.E., & Stevens, C.D. (1959) The regression analysis of causal paths. *Biometrics*, 15, 236–258.

Warren, R.D., White, J.K., & Fuller, W.A. (1974) An errors in variables analysis of managerial role performance. *Journal of the American Statistical Association*, 69, 886–893.

Wehrle, G. (1982) *Alienation in work and leasure among secretaries*. State University of New York at Stony Brook: Doctoral Dissertation.

Werts, C.E., & Linn, R.L. (1970) Path analysis: Psychological examples. *Psychological Bulletin*, 67, 193–212.

Werts, C.E., Rock, D.A., Linn, R.L., & Jöreskog, K.G. (1976) A comparison of correlations, variances, covariances and regression weights with or without measurement errors. *Psychological Bulletin*, 83, 1007–1013.

Werts, C.E., Rock, D.A., Linn, R.L., & Jöreskog, K.G. (1977) Validating psychometric assumptions within and between populations. *Educational and Psychological Measurement*, 37, 863–871.

Wheaton, B., Muthén, B., Alwin, D., & Summers, G. (1977) Assessing reliability and stability in panel models. In D.R. Heise (Ed.): *Sociological Methodology 1977*. San Francisco: Jossey-Bass.

Wiley, D.E., & Hornik, R. (1973) Measurement error and the analysis of panel data. *Mehrlicht! Studies of Educative Processes*, Report No. 5, 1973. University of Chicago.

Wright, S. (1934) The method of path coefficients. *Annals of Mathematical Statistics*, 5, 161–215.

Author Index

Aish, 45, 46, 96, 113
Akaike, 119
Alwin, 51

Bagozzi, 34
Bentler, 82, 125
Bielby, 114
Blalock, 11
Bohrnstedt, 29
Bollen, 12, 19, 111, 114, 122, 125,
    126, 155, 200, 209
Bonett, 125
Bozdogan, 119
Brett, 125
Browne, 117, 119, 122, 123, 124,
    129, 131, 183, 184

Calsyn, 16
Clelland, 12
Costner, 112
Cudeck, 39, 117, 119, 123, 124,
    129, 131, 183

Draper, 2
Duncan, 11, 29, 38, 39

Finn, 6
French, 22
Fuller, 188

Goldberger, 2, 11

Guilford, 22

Haller, 38, 39
Hauser, 11
Heise, 29
Hershberger, 114
Hodge, 199
Hoelter, 126
Holzinger, 23, 71
Hornik, 192
Huba, 122
Huitema, 6

Jackson, 51
Jagodzinski, 29
James, 125
Johnston, 2
Judd, 112
Jöreskog, 6, 23, 29, 30, 34, 45, 46,
    51, 52, 96, 111, 112, 113,
    114, 116, 122, 127, 133,
    153, 166, 180, 181, 184

Kaplan, 127, 147
Kenny, 16, 112
Kerckhoff, 187
Kluegel, 201, 203
Kühnel, 29

Lee, 114

Lewis, 125
Liang, 126
Linn, 11, 51, 52
Lomax, 51
Long, 111

Magidson, 79, 82
Maiti, 122
Mare, 51, 56, 60
Mason, 51, 56, 60
Matsueda, 114
McDonald, 12, 123, 196
McGaw, 51
Meredith, 75
Mukherjee, 122
Mulaik, 125
Muthén, 112

Olsson, 112

Portes, 38, 39

Rock, 51, 52, 190

Saris, 114, 127
Satorra, 114, 127
Singleton, 203
Smith, 2
Sobel, 155
Starnes, 203
Steiger, 124
Stelzl, 114
Stevens, 11
Swineford, 23, 71
Sörbom, 6, 29, 30, 34, 45, 46, 51,
        61, 71, 79, 112, 114, 116,
        122, 127, 133, 147, 153,
        166, 180, 181

Tanaka, 122
Theil, 11
Thurstone, 22, 196
Treiman, 199
Tucker, 125
Turner, 11

Warren, 188
Wehrle, 193
Werts, 11, 51, 52
Wheaton, 29
White, 188
Wiley, 192
Woodward, 82
Wright, 11

Subject Index

A (for **A**ll Modification Indices), 93, 108
A priori specified models, 118
AD, 182
Adding an error covariance, 96
Adding or deleting a path, 94
Admissibility Check, 179, 181, 182
AGFI, 123
AIC measure, 119
Alternative hypothesis, 118
Alternative models, 115, 119
Assessment of model fit, 23
Asymptotic covariance matrix, 45
    from file, 170
Asymptotic variances and covariances, 116
    from file, 169, 170
Attenuated correlations, 20

B (for **B**asic Model), 87, 108
B-diagram, 87
BETA(β), 137
Biserial correlation, 45
Bivariate regression, 6

CAIC, 119
Calibration sample, 129
Cause variables, 11
Censored variables, 45

CFI, 125
Chi-square, 121
Classification of variables, 136
Command line, 161
Completely standardized solution, 152
Confidence interval, 19, 118
Confirmatory factor analysis, 22, 23
Constant term, 5
Covariance Matrix, 161
    from file, 161
Covariance structure, 115
Cross-sectional studies, 113
Cycle, 154

D (**D**isplay the output file), 94, 108
Data file, 73
Degrees of freedom, 123
Delta(δ)-variables, 136
Dependent variable y, 1
Dependent variables, 15
Diagonally weighted least squares, 116, 180
Direct causal contribution, 11
Direct effect, 154
Directly observed variables, 28
Disattenuated correlation, 19
DWLS, 116, 181

E (for Parameter Estimates), 86, 108
ECVI, 119
EF (Print total and indirect effects), 184
Effect variable, 11
Empty path diagram, 100
End of Problem, 162, 185
EPC, 127
Epsilon(ϵ)-variables, 136
Equal error variances, 177
Equal factor structures, 52
Equal paths, 177
Equal regressions, 61
Equality constraints, 42, 177
Equations, 171
Error covariance, 96, 104
Error term, 1
Errors in equations, 20
Errors in variables, 20
Estimation of means of latent variables, 66
Eta(η)-variables, 88, 136
Example 1, 2, 86
Example 2, 6
Example 3, 12, 100, 105
Example 4, 16, 101, 102, 104
Example 5, 23, 92
Example 6, 29, 87, 103
Example 7, 34
Example 8, 38
Example 9, 45, 183
Example 10, 52
Example 11, 56
Example 12, 61
Example 13, 66
Example 14, 71

Example 15, 77
Example 16, 79
Example 17, 134, 159
Exploratory analysis, 22

F (for Fit Measures), 92, 108
F1 Help, 109
F3 Reestimate the model, 109
Factor analysis, 23
Factor loadings, 19
Fallible measure, 37
Fit function, 117
Fit statistics, 92
Fitted and Standardized Residuals, 146
Fitted residuals, 126
Fitting and testing, 115
Fixed, 1
Format line, 73
Free, 178
FROM-variable, 172
FS (Print factor scores regression), 184
F-statistic, 5
FULL, 168

G (for Goodness-of-Fit Measures), 92
GAMMA(γ), 137
Generalized least squares, 116, 180
Generally weighted least squares, 116, 180
Geometric series, 154
GLS, 116, 180
Goodness-of-fit index (GFI), 123
Goodness-of-fit indices, 122

Goodness-of-fit statistics, 10, 129, 144

Hypothetical constructs, 111
Hypothetical Model, 133, 147

IFI, 125
IN, 106
Independent variables, 15
Independent, 1
Indicators, 15
Indirect effect, 154
Input file, 161
INPUT.PDM, 106
Instrumental variables, 116, 180
Intercept term, 5
IT, 182
Iterations, 162, 179, 182
IV, 116, 184

Ksi(ξ)-variables, 88, 136

Labels, 72
    from file , 163
LAMBDA-X($\lambda^{(x)}$), 137
LAMBDA-Y($\lambda^{(y)}$), 137
Latent variable, 15, 18, 28, 171
Left-hand variable, 171
Let Path, 177
Let the Error Variance, 175
Let the Errors between, 176
Let, 177
Likelihood ratio test statistic, 118
LISREL
    estimates, 141
    format, 133
    input language, 133
    output, 162, 174, 184
Longitudinal research design, 29

LX, 137
LY, 137

M (for Modification Indices), 93, 108
Maximum likelihood, 48, 116, 180
Maximum likelihood solution, 75
Maximum number of iterations, 182
M-diagram, 93, 94
Means, 35, 169
Measurement errors, 16, 23, 38, 175
Measurement model, 15, 16
    for the y-variables, 90
Measures, 144
Memory, 29
Method, 162
Method of estimation, 179
Method:
    Generalized Least-Squares, 181
    Instrumental Variables Method, 181
    Two-Stage Least-Squares, 181
    Unweighted Least-Squares, 181
mi=5, 108
Missing values, 3
ML, 116, 180
Model assessment, 120
Model generating, 115, 120, 128
Model modification index, 23
Modification, 120
Modification index, 26, 93, 107, 127, 147

MR (Equivalent to RS and EF and VA), 184
Multiple correlation, R^2, 5, 20
Multitrait-multimethod, 23

N (Next group), 108
Nature of inference, 114
NFI, 125
NNFI, 125
Non-admissible solution, 182
Non-central chi-square distribution, 118
Non-centrality parameter, 118
Non-recursive models, 154
Non-recursive system, 38
Null hypothesis, 118
Number of decimals, 162, 179, 180

Observed Variables, 18, 35, 162
    from file, 163
Omitted variable bias, 113
Omitted variables bias, 2
Options: GL, 181
Options: IV, 181
Options: TS, 181
Options: UL, 181
Options, 162, 178
Ordered categories, 44
Ordinal variable, 44, 45
Output file, 133

P (Print the path diagram), 109
Panel models, 113
Panel study, 29
Parameter matrices, 137
Parameter specifications, 140

Path analysis, 11, 28
    with latent variables, 28
Path diagram, 7, 24, 162
    DOS command for **T**-values, 86
    with parameter estimates, 86
PATHDIAG, 107
PC (Print correlations of parameter estimates), 184
PDF, 123
PGFI, 125
PHI(Φ), 139
Physical line, 161
PNFI, 125
Political attitudes, 45
Polychoric correlation, 45
    coefficient, 44
Polyserial correlation coefficient, 44
Population discrepancy function, 123
PRELIS, 3
Print residuals, 179
Print completely standardized solution, 184
Printing a path diagram, 105
Product-moment correlation, 35
PSI(Ψ), 139
PT (Print technical information), 184

Q (Quit looking at path diagrams), 86, 109
Q-plot, 127, 147

R (for ERror Covariances), 87, 90, 108
Random disturbance terms, 11

Raw data from file, 166
R-diagram, 90
Reciprocal causation, 154
Recursive system, 12, 29
Recursive, 28
Reduced form, 11
Reference variable, 173
Regression coefficients, 11
Regression equation, 1
Regression model, 1, 5
    with latent variables, 77
Relations, 171
Relationships, 162, 171
Relaxing an equality constraint,
    97
Reliability, 16, 37, 60
Residual, 146
Retest effects, 29
RFI, 125
Right-hand variables, 171
RMSEA, 124
RS, 179

S (for Structural
    Relationships), 87, 108
Sample size, 161
Save sigma in file, 179, 183
Saving the Path Diagram, 106
Scale-invariant, 5
SC, 115, 152, 157, 184
S-diagram, 89
SE Print standard errors of total
    and indirect effects, 189
Selection of variables, 11, 170
Semicolon, 161
SEPC, 127
Set
    Correlations, 176

Covariances, 176
    Error Covariance, 176
    Error Variance, 175
    Path, 177
SI, 183
Significance levels, 107
    for modification indices, 108
    for t-values, 107
Significant digits, 180
SIMPLIS format, 133
SIze=sx,sy, 106
SMC, 121
Specific factor, 23
Squared multiple correlation, 5,
    60, 121
SS, 152, 157
SS (Print standardized solution),
    184
Stability index, 155
Standard deviations, 35, 169
Standard errors, 4, 37
Standardized, 19
Standardized completely, 152
Standardized residual, 126, 146
Standardized solutions, 151, 158
Starting values, 174
Statistical model, 112
Stemleaf plot, 127
Strategy analysis, 128
Strictly confirmatory, 115
Structural equation, 11
Structural equation model, 34,
    89, 111
Structural equation system, 28
Structural errors, 175
Structural parameters, 11
Symmetric, 168

System of equations, 12

T (*t*-values), 108
Tetrachoric correlation
    coefficient, 45
THETA-DELTA(Θ_δ), 139
THETA-DELTA-EPS($\Theta_{\delta\epsilon}$), 139
THETA-EPS(Θ_ϵ), 139
Threshold, 44
Title line, 162
TO-variable, 172
Total effect, 154
TSLS, 116, 180
tv=1, 107
t-value, 4, 107
Two-stage least squares, 116, 180

UL=x,y, 106
ULS, 116, 180
Unobserved variable, 171
Unweighted least squares, 116,
    180

VA (Print variances and
    covariances), 184
Validation sample, 129
Validity, 16
VarA $\to$ VarB, 178
VarB = VarA, 178
Variables, 1, 19
    selection of, 11, 170

Weight matrix, 45
Weighted least squares, 48, 116,
    180
Wide Print, 179
WLS method, 45
WLS, 116, 180
WP, 179

X (for X-Measurement
    Relationships), 87, 89, 108
X-diagram, 89
X-variables, 136

Y (for Y-Measurement
    Relationships), 87, 90, 108
Y-diagram, 90
Y-variables, 136

Z (for Zoom), 104, 108
Zeta(ζ)-variables, 136
Zooming, 104